LONDON MATHEMATICAL SOCIETY STUDENT TEXTS

Managing editor: Professor C.M. Series, Mathematics Institute
University of Warwick, Coventry CV4 7AL, United Kingdom

3   Local fields, J.W.S. CASSELS
4   An introduction to twistor theory: Second edition, S.A. HUGGETT & K.P. TOD
5   Introduction to general relativity, L.P. HUGHSTON & K.P. TOD
7   The theory of evolution and dynamical systems, J. HOFBAUER & K. SIGMUND
8   Summing and nuclear norms in Banach space theory, G.J.O. JAMESON
9   Automorphisms of surfaces after Nielsen and Thurston, A. CASSON & S. BLEILER
11 Spacetime and singularities, G. NABER
12 Undergraduate algebraic geometry, MILES REID
13 An introduction to Hankel operators, J.R. PARTINGTON
15 Presentations of groups: Second edition, D.L. JOHNSON
17 Aspects of quantum field theory in curved spacetime, S.A. FULLING
18 Braids and coverings: selected topics, VAGN LUNDSGAARD HANSEN
20 Communication theory, C.M. GOLDIE & R.G.E. PINCH
21 Representations of finite groups of Lie type, FRANÇOIS DIGNE & JEAN MICHEL
22 Designs, graphs, codes, and their links, P.J. CAMERON & J.H. VAN LINT
23 Complex algebraic curves, FRANCES KIRWAN
24 Lectures on elliptic curves, J.W.S. CASSELS
26 An introduction to the theory of L-functions and Eisenstein series, H. HIDA
27 Hilbert Space: compact operators and the trace theorem, J.R. RETHERFORD
28 Potential theory in the complex plane, T. RANSFORD
29 Undergraduate commutative algebra, M. REID
31 The Laplacian on a Riemannian manifold, S. ROSENBERG
32 Lectures on Lie groups and Lie algebras, R. CARTER, G. SEGAL & I. MACDONALD
33 A primer of algebraic D-modules, S.C. COUTINHO
34 Complex algebraic surfaces, A. BEAUVILLE
35 Young tableaux, W. FULTON
37 A mathematical introduction to wavelets, P. WOJTASZCZYK
38 Harmonic maps, loop groups and integrable systems, M. GUEST
39 Set theory for the working mathematician, K. CIESIELSKI
40 Ergodic theory and dynamical systems, M. POLLICOTT & M. YURI
41 The algorithmic resolution of diophantine equations, N.P. SMART
42 Equilibrium states in ergodic theory, G. KELLER
43 Fourier analysis on finite groups and applications, A. TERRAS
44 Classical invariant theory, P. OLVER
45 Permutation groups, P.J. CAMERON
47 Introductory lectures on rings and modules, J. BEACHY
48 Set theory, A. HAJNÁL & P. HAMBURGER
49 An introduction to K-theory for C^*-algebras, M. RØRDAM, F. LARSEN & N. LAUSTSEN
50 A brief guide to algebraic number theory, H.P.F. SWINNERTON-DYER
51 Steps in commutative algebra: Second edition, R.Y. SHARP

London Mathematical Society Student Texts 50

A Brief Guide to Algebraic Number Theory

H. P. F. Swinnerton-Dyer
University of Cambridge

CAMBRIDGE
UNIVERSITY PRESS

PUBLISHED BY THE PRESS SYNDICATE OF THE UNIVERSITY OF CAMBRIDGE
The Pitt Building, Trumpington Street, Cambridge, United Kingdom

CAMBRIDGE UNIVERSITY PRESS
The Edinburgh Building, Cambridge CB2 2RU, UK
40 West 20th Street, New York, NY 10011–4211, USA
477 Williamstown Road, Port Melbourne, VIC 3207, Australia
Ruiz de Alarcón 13, 28014 Madrid, Spain
Dock House, The Waterfront, Cape Town 8001, South Africa

http://www.cambridge.org

First published 2001
Reprinted 2002

Printed in the United Kingdom at the University Press, Cambridge

A catalogue record for this book is available from the British Library

ISBN 0 521 80292 X hardback
ISBN 0 521 00423 3 paperback

Contents

Preface   *page* vii

1   Numbers and Ideals   1
  1   The ring of integers   1
  2   Ideals and factorization   9
  3   Embedding in the complex numbers   18
  4   Change of fields   23
  5   Normal extensions   26

2   Valuations   31
  6   Valuations and completions   31
  7   Field extensions and ramification   40
  8   The Different   43
  9   Idèles and Adèles   48

3   Special fields   55
  10   Quadratic fields   55
  11   Pure cubic fields   62
  12   Biquadratic fields   63
  13   Cyclotomic fields   65
  13.1   Class numbers of cyclotomic fields   68
  13.2   Fermat's Last Theorem   73

4   Analytic methods   79
  14   Zeta functions and *L*-series   79
  15   Analytic continuation and the functional equation   84
  16   Density theorems   94

5   Class Field Theory   98
  17   The classical theory   98
  18   Chevalley's reformulation   103

19   Reciprocity theorems   106
20   The Kronecker-Weber Theorem   112
Appendix   117
A1   Prerequisites   117
A1.1 Finitely generated abelian groups and lattices   117
A1.2 Norms and Traces   121
A1.3 Haar measure   122
A2   Additional topics   124
A2.1 Characters and duality   125
A2.2 Fourier transforms   128
A2.3 Galois theory for infinite extensions   132
Exercises   135
Suggested further reading   143
Index   145

Preface

This book is intended both for number theorists and more generally for working algebraists, though some sections (notably §15) are likely to be of interest only to the former. It is largely an account of mainstream theory; but for example Chapter 3 and §20 should be seen as illustrative applications.

An **algebraic number field** is by definition a finite extension of **Q**, and algebraic number theory was initially defined as the study of the properties of algebraic number fields. Like any empire, its borders have subsequently grown. The higher reaches of algebraic number theory are now one of the crown jewels of mathematics. But algebraic number theory is not merely interesting in itself. It has become an important tool over a wide range of pure mathematics; and many of the ideas involved generalize, for example to algebraic geometry. Some applications to Diophantine equations can be found among the exercises, but there has not been room for other applications.

Algebraic number theory was originally developed to attack Fermat's Last Theorem — the assertion that $X^n + Y^n = Z^n$ has no non-trivial integer solutions for $n > 2$. It provided proofs that many values of n are impossible; some of the simpler arguments are in §13. But it did not provide a proof for all n, though recently the theorem has been proved by Andrew Wiles, assisted by Richard Taylor, by much more sophisticated methods (which still use a great deal of algebraic number theory). There are still respectable mathematicians seeking a more elementary proof, and this is not a ridiculous quest; but even if a more elementary proof is found, it is almost bound to be highly sophisticated.

There are two obvious ways of approaching algebraic number theory, one by means of ideals and the other by means of valuations. Each has its advantages, and it is desirable to be familiar with both. They are covered

in Chapters 1 and 2 respectively. In this book I have chosen to put the
main emphasis on ideals, but properties which really relate to local fields
(whether or not the latter are made explicit) are usually best handled
by means of valuations. Chapter 3 then applies the general theory to
particular kinds of number field. The first two chapters (perhaps omitting
§9), together with the easier parts of Chapter 3 and the first half of the
Appendix, would form a satisfactory and self-contained one-term graduate
course.

Though §9 is more advanced than the rest of Chapter 2, its logical home
is there; it is needed in Chapters 4 and 5, and introduces language which is
widely used across number theory. The somewhat peripheral §12 depends
on the results stated in §14 and proved in §15 of Chapter 4, as do parts
of §13.1, and thus they are not in the correct logical order; but there are
advantages in collecting all the information on special kinds of number field
in a single chapter.

There are important results which, though not in appearance analytic,
can as far as we know only be proved by analytic methods. Indeed it has
been said: 'The zeta function knows everything about the number field; we
just have to prevail on it to tell us.' Some of what it has already told us
can be found in Chapter 4.

The more advanced parts of the algebraic theory are generally known as
class field theory; most of the proofs involve Galois cohomology, either
openly or in disguise. Anyone who writes a book on algebraic number
theory is faced with a dilemma when he comes to class field theory. Most
authors stop short of it; but working algebraists ought to know the main
results of class field theory, though few of them need to understand the
rather convoluted proofs. I would think it wrong to make no mention
of class field theory; but to have included the proofs and the necessary
background material would have doubled the length of the book without
doubling its value. In consequence §§17 and 18 present an exposition of
class field theory without proofs. In §19 we deduce the general reciprocity
theorems, which are the simplest major applications of class field theory. In
addition, §20 contains a proof of the Kronecker-Weber Theorem that every
abelian extension of the rationals is cyclotomic; it is this result which made
the general structure of classical class field theory plausible long before it
was proved. The proof of the Kronecker-Weber Theorem is also rather
convoluted, but it illustrates most of the ideas in the first two chapters.

The reader needs to know the standard results about field extensions
of finite degree, including the relevant Galois theory. The properties of
finitely generated abelian groups and lattices, and of norms and traces, are

described in §A1.1 and §A1.2; most readers will already know these results, but those who do not will need to start by reading these two subsections. The existence of Haar measure and the Haar integral (described without proofs in §A1.3) is a fact which all working mathematicians should know, though again they have no need to study the proofs. Indeed, the main use of the general theory is to provide motivation and guidance; in any particular case one can expect to be able to define explicitly an integral having the required properties, and thereby evade any appeal to the general theory. The status of §A2 is rather different. The Galois theory of infinite extensions is not actually needed anywhere in this book; but anyone who uses the results in Chapter 5 may need to consider field extensions of infinite degree. The remaining subsections of §A2 cover (without proofs) characters, duality and Fourier transforms on locally compact abelian groups; these are prerequisites for §15, but also for much of advanced number theory.

The book concludes with a substantial collection of exercises. Others can be found in the text; see the index. The latter are results which are too peripheral to justify the provision of a detailed proof but which may be interesting or useful to the reader. Each of them is provided with 'stepping-stones': intermediate results which are individually not too difficult and which should enable the reader to construct a complete proof.

1

Numbers and Ideals

1 The ring of integers

Unless otherwise stated, throughout this book K and k will be algebraic number fields, even though some results hold more generally. The ring of integers of k (yet to be defined) will be denoted by $\mathfrak{o}$ or $\mathfrak{o}_k$; the ring of integers of K will be $\mathfrak{O}$ or $\mathfrak{O}_K$. What are the properties which one would like the integers of k to have? Some obvious ones are the following:-

1. $\mathfrak{o}_k$ is a commutative ring.
2. $\mathfrak{o}_k \cap \mathbf{Q} = \mathbf{Z}$, so that the integers which are rational are just the rational integers.
3. $\mathfrak{o}_k \otimes_{\mathbf{Z}} \mathbf{Q} = k$, so that each α in k can be written as $c\beta$ where c is in $\mathbf{Q}$ and β is an integer in k.
4. If α is in $\overline{\mathbf{Q}}$, the algebraic closure of $\mathbf{Q}$, the property that α is an integer only depends on α and not on the field in which we are working; in other words,

$$\mathfrak{o}_k = k \cap \{\text{integers of } \overline{\mathbf{Q}}\}.$$

5. If α and α' are conjugate over $\mathbf{Q}$ and α is an integer, then so is α'.

There is a largest subring of k satisfying these conditions, but no smallest one; so we shall choose $\mathfrak{o}_k$ to be the largest such subring. It follows from 1, 2 and 5 that if α is an integer then its monic irreducible polynomial over $\mathbf{Q}$ has coefficients in $\mathbf{Z}$. We shall say that α in $\overline{\mathbf{Q}}$ is an **algebraic integer** if it satisfies one of the three equivalent conditions in the following theorem.

Theorem 1 *Let α be an element of $\overline{\mathbf{Q}}$; then the following conditions on α are equivalent:*

(i) $\mathbf{Z}[\alpha]$ *is a finitely generated $\mathbf{Z}$-module;*

(ii) α *is a root of a monic polynomial with coefficients in* $\mathbf{Z}$*;*

(iii) *the monic minimal polynomial for* α *over* $\mathbf{Q}$ *has coefficients in* $\mathbf{Z}$*.*

Proof It is trivial that (iii)$\Rightarrow$(ii). To prove (ii)$\Rightarrow$(i), let m be the degree of the monic polynomial given by (ii); then α^m is in the $\mathbf{Z}$-module generated by $1, \alpha, \dots, \alpha^{m-1}$, whence α^μ is in the same module for all $\mu > m$. To prove (i)$\Rightarrow$(ii) choose elements $f_1(\alpha), \dots, f_n(\alpha)$ which span the $\mathbf{Z}$-module $\mathbf{Z}[\alpha]$, where the $f_\nu(X)$ are in $\mathbf{Z}[X]$. For any $N > 0$ there is a relation

$$\alpha^N - \sum b_\nu f_\nu(\alpha) = 0$$

where the b_ν are in $\mathbf{Z}$, and if N is greater than the degree of any of the f_ν this is the monic equation required.

It remains to prove (ii)$\Rightarrow$(iii). Let $f(X)$ with coefficients in $\mathbf{Z}$ be the monic polynomial for α given by (ii), and let $g(X)$ be the monic minimal polynomial for α over $\mathbf{Q}$. There is a monic polynomial $h(X)$ with coefficients in $\mathbf{Q}$ such that $f(X) = g(X)h(X)$. If (iii) were false then there would be a prime p which divided the denominator of at least one of the coefficients of $g(X)$. Let p^u with $u > 0$ be the greatest power of p which divides any of the denominators of the coefficients of $g(X)$ and let p^v with $v \geqslant 0$ be the greatest power of p which divides any of the denominators of the coefficients of $h(X)$. Let $\mathbf{F}_p$ denote the finite field of p elements. In

$$p^{u+v}f(X) = \{p^u g(X)\}\{p^v h(X)\}$$

all the coefficients have denominators prime to p; reducing mod p we obtain an expression for 0 as the product of two non-zero polynomials in $\mathbf{F}_p[X]$. This is impossible; hence (iii) must be true. $\square$

With this definition requirements 2 to 5 are trivial. To prove 1, let α, β be algebraic integers, so that there are corresponding polynomials satisfying (ii) with degrees m, n respectively. Thus $1, \alpha, \dots, \alpha^{m-1}$ span the $\mathbf{Z}$-module $\mathbf{Z}[\alpha]$ and similarly for β; hence the $\alpha^\mu \beta^\nu$ with $0 \leqslant \mu < m$, $0 \leqslant \nu < n$ span the $\mathbf{Z}$-module $\mathbf{Z}[\alpha, \beta]$. Since it is finitely generated, so are its submodules $\mathbf{Z}[\alpha \pm \beta]$ and $\mathbf{Z}[\alpha\beta]$; so $\alpha \pm \beta$ and $\alpha\beta$ are integers.

Lemma 1 *If k is identified with $\mathbf{Q}^n$ where $[k : \mathbf{Q}] = n$, then $\mathfrak{o}_k$ is a lattice in k.*

Proof Let $\alpha_1, \dots, \alpha_n$ be a base for k as a $\mathbf{Q}$-vector space. We can find $m_\nu \neq 0$ in $\mathbf{Z}$ such that the $\beta_\nu = m_\nu \alpha_\nu$ are in $\mathfrak{o}_k$; so $\mathfrak{o}_k$ spans k. For any ℓ_ν in $\mathbf{Q}$ write $\xi = \sum \ell_\nu \alpha_\nu$. If $\mathfrak{o}_k$ is not discrete in k then there are arbitrarily small $\ell_1, \dots, \ell_n$ such that ξ is non-zero and in $\mathfrak{o}_k$; hence $\text{norm}_{k/\mathbf{Q}}\xi$ is in

Z, and it is non-zero because it is the product of conjugates of ξ. But $\text{norm}\,\xi = \phi(\ell_1,\ldots,\ell_n)$ where ϕ is a homogeneous polynomial of degree n with coefficients in **Q**; so we can choose the ℓ_ν so small that $|\phi| < 1$. $\square$

More generally, an **order** in k is defined to be any subring of o_k which contains 1 and has finite index in o_k considered as a **Z**-module. Equivalently, an order is a subring of o_k which satisfies the analogues of 1, 2 and 3. It is easy to write down some orders in k; for example, if $n = [k : \mathbf{Q}]$ and α is any integer in k such that $k = \mathbf{Q}(\alpha)$, then the **Z**-module spanned by $1,\alpha,\ldots,\alpha^{n-1}$ is an order in k.

Let $R_1 \subset R_2$ be commutative rings with a 1. We shall say that R_1 is **integrally closed** in R_2 if α in R_2, all β_ν in R_1 and

$$\alpha^n + \beta_1\alpha^{n-1} + \cdots + \beta_n = 0$$

together imply that α is in R_1.

Lemma 2 *The ring of integers in $\overline{\mathbf{Q}}$ is integrally closed in $\overline{\mathbf{Q}}$.*

Proof We argue as in the proof of 1 on page 1. In the notation above, the **Z**-module $\mathbf{Z}[\beta_1,\ldots,\beta_n]$ is finitely generated because it is contained in the ring of integers of $\mathbf{Q}(\beta_1,\ldots,\beta_n)$; suppose the finite set S spans it. In an obvious notation $S, S\alpha,\ldots,S\alpha^{n-1}$ span $\mathbf{Z}[\alpha,\beta_1,\ldots,\beta_n]$; hence the latter is finitely generated and so is its submodule $\mathbf{Z}[\alpha]$. $\square$

The identification of k with $\mathbf{Q}^n$ which we used in Lemma 1 does not enable us to compute a meaningful measure for k/o_k, because there is no natural measure on the $\mathbf{Q}^n$ in Lemma 1 which is not derived from o_k. But there is an identification of $k \otimes_{\mathbf{Q}} \mathbf{R}$ which does achieve this. Let σ_ν run through the n embeddings $k \to \mathbf{C}$. Of these, there are r_1 embeddings into $\mathbf{R}$ and r_2 pairs of complex conjugate embeddings whose images are not in $\mathbf{R}$. Here $r_1 + 2r_2 = n$; and r_1, r_2 will almost always have these meanings. Denote temporarily by $\sigma : k \to \mathbf{C}^n$ the map given by $\alpha \mapsto (\sigma_1\alpha,\ldots,\sigma_n\alpha)$ and by V the **R**-vector space $\sigma k \otimes_{\mathbf{Q}} \mathbf{R}$; since $\sigma_\nu k \subset \mathbf{R}$ if σ_ν is real, and $(\sigma_\nu \times \overline{\sigma_\nu})k$ is contained in a space $\mathbf{R}^2$ if $\sigma_\nu, \overline{\sigma_\nu}$ are complex conjugate, $\dim V \leqslant n$. But $o_k \to \sigma o_k$ is an isomorphism because it has trivial kernel, so σo_k is a free **Z**-module on n generators; and σo_k is discrete in $\mathbf{C}^n$ and hence in V, for the same reason as in the proof of Lemma 1. So σo_k is a lattice in V, and $\dim V = n$.

The inclusion $V \subset \mathbf{C}^n$ induces a canonical measure of volume in V. The measure of $V/\sigma o$ is finite and non-zero; it tells one how sparse the integers of k are. With a suitable choice of the canonical measure on V, the volume of $V/\sigma\mathfrak{o}$ is $|\det M|$, where $\alpha_1,\dots,\alpha_n$ are a base for $\mathfrak{o}_k$ and M is the matrix of the $\sigma_\mu\alpha_\nu$. Unfortunately $\det M$ is only defined up to sign, and is not necessarily in either **R** or k; so instead we consider

$$d_k = (\det(M))^2 = \det(\,^t M M) = \det(\mathrm{Tr}_{k/\mathbf{Q}}\,(\alpha_\mu\alpha_\nu)).$$

This is called the **absolute discriminant** of k. Observe that M has r_2 pairs of complex conjugate rows, so its determinant is i^{r_2} times a real number; thus the sign of d_k is $(-1)^{r_2}$.

More generally, let $\beta_1,\dots,\beta_n$ in $\mathfrak{o}_k$ be linearly independent over **Z**, and let m be the index in $\mathfrak{o}_k$ of the **Z**-module generated by $\beta_1,\dots,\beta_n$. If

$$\Delta^2(\beta_1,\dots,\beta_n) = \det(\mathrm{Tr}(\beta_\mu\beta_\nu)) \tag{1}$$

then a comparison of the last two displayed equations gives $\Delta^2 = m^2 d_k$. Conversely, if we start from linearly independent integers $\beta_1,\dots,\beta_n$ and try to find a base for $\mathfrak{o}_k$, the fact that d_k is an integer restricts us to finitely many possible m.

We can further generalize (1) by considering $K \supset k$ with $[K : k] = n$. Let $\alpha_1,\dots,\alpha_n$ be a base for K as a k-vector space, and write

$$\Delta^2_{K/k}(\alpha_1,\dots,\alpha_n) = \det(\mathrm{Tr}_{K/k}(\alpha_\mu\alpha_\nu)).$$

Then $\Delta^2_{K/k}$ is non-zero. For otherwise there would be a non-zero column vector $\,^t(\xi_1,\dots,\xi_n)$ killed by the matrix $(\mathrm{Tr}_{K/k}(\alpha_\mu\alpha_\nu))$, where the ξ_ν are in k. Set $\gamma = \sum \xi_\nu\alpha_\nu$; then $\mathrm{Tr}(\alpha_\mu\gamma) = 0$ for each μ and therefore $\mathrm{Tr}(\alpha\gamma) = 0$ for each α in K. But this cannot be true for $\alpha = \gamma^{-1}$. This argument, and the two lemmas which follow, are valid for any field k of characteristic 0.

The following result will be needed in §8.

Lemma 3 *Let K, k be algebraic number fields with $K \subset k$; then every k-linear map $K \to k$ is given by $\alpha \mapsto \mathrm{Tr}_{K/k}(\beta\alpha)$ for some β in K.*

Proof Call this map ϕ_β. The k-linear map from K to the dual space of K given by $\beta \mapsto \phi_\beta$ has trivial kernel. The two spaces involved have the same dimension as k-vector spaces, so the map is an isomorphism. $\square$

The following result, which will be needed in §13.1, is known as Hilbert's Theorem 90; in highbrow language it states that a certain cohomology group is trivial.

Lemma 4 *Let K/k be a Galois extension whose Galois group $\mathrm{Gal}(K/k)$ is cyclic with generator σ. If α in K is such that $\mathrm{norm}_{K/k}\alpha = 1$ then $\alpha = \beta/\sigma\beta$ for some β in K; and we can take β to be integral.*

Proof Let $[K : k] = n$ and for any γ in K consider

$$\beta = \gamma \cdot \alpha + \sigma\gamma \cdot \alpha \cdot \sigma\alpha + \cdots + (\sigma^{n-1}\gamma)(\alpha \cdots \sigma^{n-1}\alpha);$$

then $\alpha \cdot \sigma\beta = \beta$. If $\beta = 0$ for every γ then $\Delta_{K/k}^2(\gamma_1, \dots, \gamma_n) = 0$ for any $\gamma_1, \dots, \gamma_n$ in K, and this we know to be false. We can make β integral by multiplying it by a suitable element of $\mathfrak{o}_k$. $\qquad\qquad$ □

One curious property of the absolute discriminant is the following.

Theorem 2 (Stickelberger) *We have* $d_k \equiv 0$ *or* $1 \bmod 4$.

Proof Write $n = [k : \mathbf{Q}]$, let $\alpha_1, \dots, \alpha_n$ be a base for $\mathfrak{o}_k$, and write

$$A = \sum_{\text{all } \pi} \left(\prod_{\mu} \sigma_\mu \alpha_{\pi(\mu)} \right), \quad B = \sum_{\pi \text{ odd}} \left(\prod_{\mu} \sigma_\mu \alpha_{\pi(\mu)} \right)$$

where π denotes a permutation of $1, \dots, n$. We have $\det M = A - 2B$ and therefore $d_k = A^2 + 4(B^2 - AB)$. Both A and B are algebraic integers, and A is rational by Galois theory; hence $B^2 - AB$ is also rational and is therefore a rational integer. Hence $d_k \equiv A^2 \equiv 0$ or $1 \bmod 4$. $\qquad$ □

The relation between k and $\mathfrak{o}_k$ which we have been discussing is a special case of a much more general one. The results which we need are not much harder to prove in the general case, provided one takes the assertions in the right (somewhat unnatural) order. Until the end of this section, we therefore consider any pair of commutative rings $R \supset \mathfrak{o}$ having a common identity element 1. (In the applications $\mathfrak{o}$ will be an integral domain and R a field; but we do not need to assume as much as this.) For any α in R consider the following three statements:

(i) $\mathfrak{o}[\alpha]$ is finitely generated as an $\mathfrak{o}$-submodule;
(i)′ $\mathfrak{o}[\alpha]$ is contained in a ring $R_\alpha \subset R$ which is finitely generated as an $\mathfrak{o}$-submodule;
(ii) α is a root of a monic polynomial with coefficients in $\mathfrak{o}$.

I claim that these statements are equivalent. Clearly (i) implies (i)′, and (ii) implies (i) as in the proof of Theorem 1. So assume that (i)′ holds and that $\beta_1, \dots, \beta_n$ span R_α as an $\mathfrak{o}$-module. We have $\alpha\beta_\mu = \sum \gamma_{\mu\nu}\beta_\nu$ for some $\gamma_{\mu\nu}$ in $\mathfrak{o}$; so α is a root of $\det(xI - \Gamma) = 0$ where Γ is the matrix of the $\gamma_{\mu\nu}$. This is just (ii).

Denote by $\mathfrak{O}$ the set of all α in R satisfying any of these conditions. An argument like that which follows the proof of Theorem 1 shows that $\mathfrak{O}$ is a ring with $o \subset \mathfrak{O} \subset R$. Similarly an argument like that in the proof of Lemma 2 shows that $\mathfrak{O}$ is integrally closed in R. If o is an integral domain, let k be its quotient field and assume $R \supset k$. In this case α is in $\mathfrak{O}$ if and only if

(iii) the monic minimal polynomial for α over k has coefficients in $\mathfrak{O} \cap k$.

Since $\mathfrak{O}$ is integrally closed in R, (iii) implies that α is in $\mathfrak{O}$. Conversely, if (ii) holds then α satisfies some monic equation $f(X) = 0$ with coefficients in o; hence so does $\sigma\alpha$ where σ is any embedding $k(\alpha) \to \bar{k}$. Thus each $\sigma\alpha$ is in the integral closure of o in $\bar{k}$. Whatever the characteristic, the monic minimal polynomial for α over k has the form

$$g(X) = \prod_{\sigma}(X - \sigma\alpha)^N$$

for some $N > 0$; and the coefficients of $g(X)$ lie both in k and in the integral closure of o in $\bar{k}$, so they lie in $\mathfrak{O} \cap k$. Note that we need not have $o = \mathfrak{O} \cap k$; if for example $o = \mathbf{Z}[\sqrt{-3}]$ and $R = \overline{\mathbf{Q}}$ then $\alpha = \frac{1}{2}(1 + \sqrt{-3})$ is in $\mathfrak{O}$ because it satisfies $\alpha^2 - \alpha + 1 = 0$, and in fact $\mathfrak{O} \cap k = \mathbf{Z}[\alpha]$.

The unnatural looking criterion (i)′ in the preceding discussion has been included because an o-submodule of a finitely generated o-module need not be finitely generated. (See Exercise 1.6.) This is frequently inconvenient; so if o is a commutative ring with a 1 we say that an o-module M is **Noetherian** if every o-submodule of M (including M itself) is finitely generated, and we say that o itself is **Noetherian** if o is a Noetherian o-module — in other words, if every ideal of o is finitely generated. Any o-submodule of a Noetherian module o is Noetherian; but a subring of a Noetherian ring need not be Noetherian.

We say that an o-module M satisfies the **ascending chain condition** if every increasing sequence $M_1 \subset M_2 \subset \cdots$ of o-submodules of M stabilizes — that is, if the sequence is eventually constant.

Lemma 5 *The three following conditions are equivalent:*

(i)   *M is Noetherian;*
(ii) *M satisfies the ascending chain condition;*
(iii) *every non-empty family of o-submodules of M contains maximal elements.*

Proof (i)⇒(ii). Suppose that $M_1 \subset M_2 \subset \cdots$ is an increasing sequence of o-submodules of M; then $M^{\sharp} = \cup M_n$ is an o-submodule of M. Hence it is generated by a finite set S of elements of $M^{\sharp}$. If M_N contains all the elements of $\mathcal{S}$, then $M^\sharp \subset M_N$; so the increasing sequence is constant from M_N on.

(ii)⇒(iii). If (iii) were false we could construct a strictly increasing sequence $M_1 \subset M_2 \subset \cdots$ of o-submodules of M. For suppose we have chosen $M_1, \ldots, M_n$. The o-submodules of M which contain M_n form a non-empty family, and this family contains no maximal elements; so we can choose an M_{n+1} which strictly contains M_n. This contradicts (ii).

(iii)⇒(i). Suppose that M contains an o-submodule N which is not finitely generated. Let $\mathcal{S}$ be the set of all finitely generated o-submodules of N, let M_0 be an element of $\mathcal{S}$ and let ξ be an element of N not in M_0. Then M_0 is not maximal in $\mathcal{S}$ because $\mathcal{S}$ contains the strictly larger o-submodule of N generated by M_0 and ξ. □

Lemma 6 *Let L, M, N be o-modules such that the sequence*

$$0 \to L \to M \to N \to 0$$

is exact. Then M is Noetherian if and only if L and N are.

Proof If M is Noetherian then L is a submodule of M and therefore also Noetherian; and if N_1 is a submodule of N then its inverse image in M is finitely generated and hence so is N_1.

Conversely, if L and N are Noetherian let $\phi : L \to M$ and $\psi : M \to N$ be the maps in the exact sequence. Let $M_1 \subset M_2 \subset \cdots$ be an increasing sequence of o-submodules of M. The increasing sequences $\{\phi(L) \cap M_\nu\}$ and $\{\psi(M_\nu)\}$ consist of o-submodules of $\phi(L) \approx L$ and N respectively, so they both stabilize — say by the n-th term. If now $\nu > n$ then

$$\phi(L) \cap M_\nu = \phi(L) \cap M_n = L^*$$

say, and $M_\nu/L^* \approx \psi(M_\nu) = \psi(M_n) \approx M_n/L^*$; so $M_\nu = M_n$ and the sequence $\{M_\nu\}$ also stabilizes by the n-th term. □

Theorem 3 *If o is Noetherian then an o-module is Noetherian if and only if it is finitely generated.*

Proof It follows from Lemma 6 by induction that the direct sum of finitely many copies of o is Noetherian. But any finitely generated o-module is a homomorphic image of such a direct sum; so it too is Noetherian. □

To study Noetherian modules over a non-Noetherian ring would be eccentric, so Theorem 3 reduces us to the study of Noetherian rings. Let $o \to o'$ be a surjective homomorphism of rings and assume that o is Noetherian.

Then $\mathfrak{o}'$ is a Noetherian $\mathfrak{o}$-module, by Theorem 3, so all its ideals are finitely generated as $\mathfrak{o}$-modules. Hence they are finitely generated as $\mathfrak{o}'$-modules; so $\mathfrak{o}'$ is Noetherian. In other words, the homomorphic image of a Noetherian ring is Noetherian.

Let $\mathfrak{O}$ be a ring containing $\mathfrak{o}$; we shall say that $\mathfrak{O}$ is finitely generated over $\mathfrak{o}$ as a ring if $\mathfrak{O} = \mathfrak{o}[\alpha_1, \dots, \alpha_n]$ for some elements $\alpha_1, \dots, \alpha_n$ in $\mathfrak{O}$.

Theorem 4 (Hilbert Basis Theorem) *If $\mathfrak{o}$ is Noetherian and $\mathfrak{O}$ is finitely generated over $\mathfrak{o}$ as a ring, then $\mathfrak{O}$ is Noetherian.*

Proof We prove first that if X is transcendental over $\mathfrak{o}$ (that is, if X is not a root of any polynomial with coefficients in $\mathfrak{o}$) then $\mathfrak{o}[X]$ is Noetherian. For let J be an ideal in $\mathfrak{o}[X]$. The set M of leading coefficients of elements of J is an $\mathfrak{o}$-ideal. Let $\mathcal{B}$ be a finite set which spans M as an $\mathfrak{o}$-module and for each element of $\mathcal{B}$ choose a polynomial in J with that element as its leading coefficient. Let $\mathcal{S}$ be the set of these polynomials, and let N be the largest of their degrees. For each n with $0 \leqslant n < N$ let M_n be the $\mathfrak{o}$-ideal of the coefficients of X^n in polynomials in J of degree at most n, and let $\mathcal{B}_n$ be a finite set which spans M_n. For each element of $\mathcal{B}_n$ choose a polynomial in J of degree at most n with that element as the coefficient of X^n, and let $\mathcal{S}_n$ be the set of these polynomials. I claim that J is spanned as an $\mathfrak{o}[X]$-module by the set $\mathcal{S}^\sharp$ which is the union of $\mathcal{S}$ and the $\mathcal{S}_n$ with $0 \leqslant n < N$. For if not, among the elements of J which do not lie in the $\mathfrak{o}[X]$-module spanned by $\mathcal{S}^\sharp$, let $f(X)$ be one of lowest degree. If $\deg f(X) \geqslant N$, let the leading coefficient of $f(X)$ be $\alpha = \sum \beta_\mu \alpha_\mu$ where the β_μ are in $\mathfrak{o}$ and the α_μ are elements of $\mathcal{B}$. Let $f_\mu(X)$ be the element of $\mathcal{S}$ whose leading coefficient is α_μ. Then

$$f(X) - \sum \beta_\mu X^{\deg f - \deg f_\mu} f_\mu(X)$$

is an element of J which has degree less than that of $f(X)$, so it is in the $\mathfrak{o}[X]$-module spanned by $\mathcal{S}^\sharp$. Hence so is f, which is a contradiction. If $\deg f(X) = n < N$ then a similar argument works, using the elements of $\mathcal{S}_n$ instead of those of $\mathcal{S}$.

If $X_1, \dots, X_m$ are independent transcendentals over $\mathfrak{o}$, it follows by induction that $\mathfrak{o}[X_1, \dots, X_m]$ is Noetherian. Finally, if $\mathfrak{O} = \mathfrak{o}[\alpha_1, \dots, \alpha_m]$ is any finitely generated ring over $\mathfrak{o}$ then it is the image of $\mathfrak{o}[X_1, \dots, X_m]$ under a suitable homomorphism; so it too is Noetherian. $\square$

If we know some Noetherian rings, Theorem 4 enables us to generate many more. But any principal ideal domain is Noetherian; in particular $\mathbf{Z}$ and all fields are Noetherian. For our purposes what matters is that $\mathfrak{o}_k$ is

Noetherian whenever k is an algebraic number field, because o_k is finitely generated over $\mathbf{Z}$ by Lemma 1.

2 Ideals and factorization

Let o_k be the ring of integers of k. Unfortunately we do not in general have unique factorization in o_k; for the standard example see Exercise 1.2. Of the various ideas that have been introduced to alleviate this situation, two have turned out to be valuable: these are ideals (described in this section) and valuations (described in Chapter 2).

Of the key theorems about ideals in o_k, Theorems 5, 6 and 7 below are usually proved in a more general setting — that of Dedekind domains. The disadvantage of this approach is that it involves one very opaque proof — in our case, that of Theorem 5. In the exercise at the end of this section the reader will find a simpler approach, but one whose validity is confined to algebraic number fields.

A **Dedekind domain** is an integral domain o with a 1 such that

(i) o is Noetherian and integrally closed in its field of fractions,
(ii) every non-zero prime ideal of o is maximal.

Denote the quotient field of o by k. The following lemma shows that we cannot get rid of non-principal ideals by replacing o by a smaller ring having the same quotient field k. We can do so by replacing o by a slightly larger ring R (see Exercise 1.10) and this is sometimes useful; but there is a price to be paid. In particular, we cannot expect R to satisfy the criteria at the beginning of §1.

Lemma 7 *Any principal ideal domain is Dedekind.*

Proof Let o be a principal ideal domain, and therefore Noetherian. Suppose that β in k is integral over o and write $\beta = \alpha_1/\alpha_2$ with α_1, α_2 in o. We can assume that $(\alpha_1, \alpha_2) = (1)$; for if $(\alpha_1, \alpha_2) = (\gamma)$ with γ not a unit, we can divide α_1 and α_2 by γ. If $\beta^n + c_1\beta^{n-1} + \cdots + c_n = 0$ where the c_ν are in o then $\alpha_1^n + c_1\alpha_1^{n-1}\alpha_2 + \cdots + c_n\alpha_2^n = 0$. It follows that $(\alpha_2) = (\alpha_1^n, \alpha_2) \supset (\alpha_1, \alpha_2)^n = (1)$, so that α_2 is a unit and β is in o. Now let (α) be a non-zero prime ideal of o and let (β) be a maximal ideal containing (α). Thus α is in (β) and hence equal to $\beta\gamma$ for some γ in o. But (α) is prime, so one of β, γ must be in (α). If β is in (α) then $(\beta) \subset (\alpha)$ so that (α) is maximal; but if $\gamma = \alpha\delta$ then $\alpha = \beta\alpha\delta$ whence $\beta\delta = 1$, and then $(\beta) = 1$ which is forbidden. $\square$

By a **fractional ideal** $\mathfrak{a}$ we mean a finitely generated $\mathfrak{o}$-submodule of k. (To avoid ambiguity, an ideal in $\mathfrak{o}$ sometimes has to be called an **integral ideal**.) If $\alpha_1, \ldots, \alpha_m$ span $\mathfrak{a}$ as an $\mathfrak{o}$-module and $\alpha_\mu = \delta_\mu / \gamma_\mu$ with γ_μ, δ_μ in $\mathfrak{o}$, then $\gamma \mathfrak{a} \subset \mathfrak{o}$ where $\gamma = \prod \gamma_\mu$. Hence any fractional ideal has the form $c\mathfrak{b}$, the set of $c\beta$ where c is a fixed element of k and β runs through the elements of some ideal $\mathfrak{b}$. Conversely, any such set $c\mathfrak{b}$ is a fractional ideal. All the obvious rules extend from ideals to fractional ideals.

Theorem 5 *The non-zero fractional ideals of a Dedekind domain form a multiplicative group.*

Proof The only difficulty is to prove the existence of inverses. The proof proceeds through a sequence of assertions.

• If $\mathfrak{a}$ is a non-zero ideal then $\mathfrak{p}_1 \mathfrak{p}_2 \cdots \mathfrak{p}_m \subset \mathfrak{a}$ for some prime ideals $\mathfrak{p}_\mu$.

Suppose the assertion is false; then because $\mathfrak{o}$ is Noetherian we can choose $\mathfrak{a}$ maximal among the ideals which do not have this property. Since $\mathfrak{a}$ is not itself prime, we can choose β_1, β_2 in $\mathfrak{o}$ but not in $\mathfrak{a}$ such that $\beta_1 \beta_2$ is in $\mathfrak{a}$. Write $\mathfrak{b}_\mu = (\mathfrak{a}, \beta_\mu)$; then each $\mathfrak{b}_\mu$ strictly contains $\mathfrak{a}$, so by maximality it contains a product of prime ideals, and hence so does $\mathfrak{a} \supset \mathfrak{b}_1 \mathfrak{b}_2$.

• Every non-zero prime ideal $\mathfrak{p}$ is invertible.

Let $\mathfrak{p}^-$ be the $\mathfrak{o}$-module of elements α in k such that $\alpha \mathfrak{p} \subset \mathfrak{o}$; we shall show that $\mathfrak{p}^-$ is the inverse of $\mathfrak{p}$ which we are looking for. Choose $\beta \neq 0$ in $\mathfrak{p}$; then $\beta \mathfrak{p}^- \subset \mathfrak{o}$ is an $\mathfrak{o}$-module and hence an ideal; so $\mathfrak{p}^-$ is a fractional ideal. Since $\mathfrak{p}^- \supset \mathfrak{o}$, we have $\mathfrak{o} \supset \mathfrak{p}^- \mathfrak{p} \supset \mathfrak{p}$; and because $\mathfrak{p}$ is maximal, we must have equality in one inclusion or the other. If we have equality in the first inclusion, $\mathfrak{p}^-$ is the inverse of $\mathfrak{p}$ which we are looking for; so we assume that $\mathfrak{p} = \mathfrak{p}^- \mathfrak{p}$ and derive a contradiction. Choose m minimal so that there exists a product

$$\mathfrak{p}_1 \mathfrak{p}_2 \cdots \mathfrak{p}_m \subset (\beta) \subset \mathfrak{p},$$

where β is as before. One of the $\mathfrak{p}_\mu$, say $\mathfrak{p}_1$, must be contained in $\mathfrak{p}$ and thus equal to $\mathfrak{p}$; for otherwise for each μ we could choose α_μ in $\mathfrak{p}_\mu$ but not in $\mathfrak{p}$, and $\prod \alpha_\mu$ would not be in $\mathfrak{p}$. By the minimality of m, $\mathfrak{p}_2 \cdots \mathfrak{p}_m \not\subset (\beta)$ and hence there exists γ in $\mathfrak{p}_2 \cdots \mathfrak{p}_m$ but not in (β). Now $\gamma \mathfrak{p} \subset (\beta)$, so that $\delta = \gamma / \beta$ is in $\mathfrak{p}^-$ but not in $\mathfrak{o}$. Thus $\delta \mathfrak{p} \subset \mathfrak{p}^- \mathfrak{p} = \mathfrak{p}$; and if $\{\beta_1, \ldots, \beta_n\}$ spans $\mathfrak{p}$ as an $\mathfrak{o}$-module then $\delta \beta_\mu = \sum \gamma_{\mu\nu} \beta_\nu$ for some $\gamma_{\mu\nu}$ in $\mathfrak{o}$. This implies $\det(\delta I - \Gamma) = 0$ where Γ is the matrix of the $\gamma_{\mu\nu}$; and since $\mathfrak{o}$ is integrally closed in k it follows that δ is an integer, which is a contradiction.

- Every non-zero integral ideal is invertible.

If not, there would be a maximal non-invertible ideal $\mathfrak{a}$. Among the ideals containing $\mathfrak{a}$ there is one which is maximal and therefore prime; denote it by $\mathfrak{p}$. Thus $\mathfrak{a} \subset \mathfrak{p}^{-1}\mathfrak{a} \subset \mathfrak{o}$. If $\mathfrak{a} = \mathfrak{p}^{-1}\mathfrak{a}$ then an argument like that for the previous displayed statement shows that $\mathfrak{p}^{-1} \subset \mathfrak{o}$, which would imply $\mathfrak{o} = \mathfrak{p}\mathfrak{p}^{-1} \subset \mathfrak{p}$. So $\mathfrak{a} \neq \mathfrak{p}^{-1}\mathfrak{a}$. By maximality $\mathfrak{p}^{-1}\mathfrak{a}$ has an inverse $\mathfrak{b}$, and $\mathfrak{b}\mathfrak{p}^{-1}$ is an inverse for $\mathfrak{a}$.

Finally, any fractional ideal $c\mathfrak{a}$ has an inverse $c^{-1}\mathfrak{a}^{-1}$. □

Corollary *For any non-zero ideals $\mathfrak{a}_1, \mathfrak{a}_2$ the four assertions $\mathfrak{a}_1 \subset \mathfrak{a}_2$, $\mathfrak{a}_1\mathfrak{a}_2^{-1} \subset \mathfrak{o}$, $\mathfrak{a}_1^{-1}\mathfrak{a}_2 \supset \mathfrak{o}$ and $\mathfrak{a}_2|\mathfrak{a}_1$ are equivalent.*

Theorem 6 *Let $\mathfrak{o}$ be a Dedekind domain. Any non-zero integral ideal $\mathfrak{a}$ in $\mathfrak{o}$ can be written as a product*

$$\mathfrak{a} = \mathfrak{p}_1 \cdots \mathfrak{p}_m \tag{2}$$

where the $\mathfrak{p}_\mu$ are prime ideals, and this expression is unique up to the order of the factors.

Proof If there are non-zero $\mathfrak{a}$ with no decomposition (2), we can assume that $\mathfrak{a}$ is maximal among them. There exists a maximal ideal $\mathfrak{p} \supset \mathfrak{a}$, so $\mathfrak{p}^{-1}\mathfrak{a}$ is an integral ideal strictly containing $\mathfrak{a}$; for $\mathfrak{p}^{-1}\mathfrak{a} = \mathfrak{a}$ would imply $\mathfrak{p}^{-1} = \mathfrak{o}$ on multiplying by $\mathfrak{a}^{-1}$. Hence $\mathfrak{p}^{-1}\mathfrak{a}$ is a product of prime ideals, and if we multiply this product by $\mathfrak{p}$ we get a product for $\mathfrak{a}$.

Suppose that there are two essentially different factorizations of some $\mathfrak{a}$. Among such ideals $\mathfrak{a}$ and their factorizations let

$$\mathfrak{a} = \mathfrak{p}_1 \cdots \mathfrak{p}_m = \mathfrak{q}_1 \cdots \mathfrak{q}_n$$

be the formula with the least value of m. We cannot have $m = 0$, because then $\mathfrak{a} = (1)$ and so also $n = 0$. If $\mathfrak{p}_1$ were not among the $\mathfrak{q}_\nu$, we could for each ν find α_ν in $\mathfrak{q}_\nu$ but not in $\mathfrak{p}_1$ and $\prod \alpha_\nu$ would be in $\mathfrak{a}$ but not in $\mathfrak{p}_1$, which gives a contradiction. So we can remove a factor $\mathfrak{p}_1$ in both expressions for $\mathfrak{a}$, and m is not minimal. □

Corollary *Let $\mathfrak{o}$ be a Dedekind domain with quotient field k. Any non-zero fractional ideal $\mathfrak{a}$ in k can be written in the form*

$$\mathfrak{a} = \mathfrak{p}_1 \cdots \mathfrak{p}_m / \mathfrak{q}_1 \cdots \mathfrak{q}_n \tag{3}$$

where the $\mathfrak{p}_\mu, \mathfrak{q}_\nu$ are prime ideals and no $\mathfrak{p}_\mu$ is equal to any $\mathfrak{q}_\nu$; and this expression is unique up to the order of the factors.

Proof We have $\mathfrak{a} = (\gamma_1/\gamma_2)\mathfrak{b}$ for some integral $\mathfrak{b}$ and some non-zero γ_1, γ_2 in $\mathfrak{o}$; applying the theorem to (γ_1), (γ_2) and $\mathfrak{b}$ gives an expression (3), and we can remove any factors common to the numerator and denominator. Suppose that there were a second such expression

$$\mathfrak{a} = \mathfrak{p}'_1 \cdots \mathfrak{p}'_{m'}/\mathfrak{q}'_1 \cdots \mathfrak{q}'_{n'};$$

then we would have

$$\mathfrak{p}_1 \cdots \mathfrak{p}_m \mathfrak{q}'_1 \cdots \mathfrak{q}'_{n'} = \mathfrak{q}_1 \cdots \mathfrak{q}_n \mathfrak{p}'_1 \cdots \mathfrak{p}'_{m'}$$

and since no $\mathfrak{p}_\mu$ is a $\mathfrak{q}_\nu$ and no $\mathfrak{p}'_\mu$ is a $\mathfrak{q}'_\nu$, it follows from the uniqueness clause in the theorem that the $\mathfrak{p}'_{\mu'}$ are a permutation of the $\mathfrak{p}_\mu$ and the $\mathfrak{q}'_{\nu'}$ a permutation of the $\mathfrak{q}_\nu$. □

The theorem which follows, and also Corollary 1 to it, are each known as the Chinese Remainder Theorem.

Theorem 7 *Let $\mathfrak{a} = \prod \mathfrak{p}_\mu^{n_\mu}$ be a non-zero ideal of $\mathfrak{o}$; then the natural map $\phi : \mathfrak{o} \to \prod(\mathfrak{o}/\mathfrak{p}_\mu^{n_\mu})$ is onto and induces an isomorphism*

$$\mathfrak{o}/\mathfrak{a} \approx \prod(\mathfrak{o}/\mathfrak{p}_\mu^{n_\mu}).$$

Proof The kernel $\mathfrak{b}$ of ϕ consists of the α which lie in each $\mathfrak{p}_\mu^{n_\mu}$; so $\mathfrak{b}$ is the largest ideal divisible by each $\mathfrak{p}_\mu^{n_\mu}$, by the Corollary to Theorem 5. Thus $\mathfrak{b} = \mathfrak{a}$. Since $\mathfrak{b}_1 = \mathfrak{p}_2^{n_2} \cdots \mathfrak{p}_m^{n_m}$ is not divisible by $\mathfrak{p}_1$ and hence is not contained in it, we can find β_1 in $\mathfrak{b}_1$ but not in $\mathfrak{p}_1$. Since $\mathfrak{p}_1$ is maximal, $\mathfrak{o}/\mathfrak{p}_1$ is a field, and hence we can find γ_1 in $\mathfrak{o}$ whose image in $\mathfrak{o}/\mathfrak{p}_1$ is the inverse of the image of β_1. Thus $\beta_1\gamma_1 = 1 - \delta_1$ with δ_1 in $\mathfrak{p}_1$. Now $1 - \delta_1^{n_1}$ is a multiple of β_1 which is congruent to $1 \bmod \mathfrak{p}_1^{n_1}$. Hence the image of ϕ contains $1 \times 0 \times \cdots \times 0$. It also contains all the other similar expressions; so ϕ is onto. □

Corollary 1 *Let $\mathfrak{a}_1, \ldots, \mathfrak{a}_m$ be non-zero integral ideals coprime in pairs and let $\alpha_1, \ldots, \alpha_m$ be elements of $\mathfrak{o}$. Then there exists α in $\mathfrak{o}$ such that*

$$\alpha \equiv \alpha_\mu \bmod \mathfrak{a}_\mu \quad (\mu = 1, \ldots, m). \tag{4}$$

Proof Write $\mathfrak{a}_\mu = \prod \mathfrak{p}_{\mu\nu}^{n_{\mu\nu}}$; by hypothesis no prime ideal occurs more than once as a $\mathfrak{p}_{\mu\nu}$, and each congruence (4) is equivalent to the finite set of congruences

$$\alpha \equiv \alpha_\mu \bmod \mathfrak{p}_{\mu\nu}^{n_{\mu\nu}}.$$

Now write $\mathfrak{a} = \prod \mathfrak{a}_\mu$; the corollary just restates that the map ϕ in the theorem is onto. $\square$

Corollary 2 *Let $\mathfrak{a}, \mathfrak{b}$ be non-zero ideals; then we can find α in $\mathfrak{a}$ such that $(\alpha)/\mathfrak{a}$ is prime to $\mathfrak{b}$.*

Proof Let $\mathfrak{p}_1, \ldots, \mathfrak{p}_m$ be the prime factors of $\mathfrak{b}$. For each $\mu = 1, \ldots, m$ define n_μ by $\mathfrak{p}_\mu^{n_\mu} \| \mathfrak{a}$, and choose α_μ in $\mathfrak{p}_\mu^{n_\mu}$ but not in $\mathfrak{p}_\mu^{n_\mu+1}$. By Corollary 1 there exists α in $\mathfrak{o}$ such that

$$\alpha \equiv \alpha_\mu \bmod \mathfrak{p}_\mu^{n_\mu+1} \text{ for each } \mu \quad \text{and} \quad \alpha \equiv 0 \bmod (\mathfrak{a}/\prod \mathfrak{p}_\mu^{n_\mu});$$

and this does what we need. $\square$

Corollary 3 *Let $\mathfrak{a}$ be a non-zero ideal and α' a non-zero element of $\mathfrak{a}$; then there exists α in $\mathfrak{a}$ such that $\mathfrak{a} = (\alpha, \alpha')$.*

Proof Apply Corollary 2 with $\mathfrak{b} = (\alpha')$. $\square$

Each of the last two corollaries extends to fractional ideals $\mathfrak{a}$. For we can find a non-zero c in $\mathfrak{o}$ such that $c\mathfrak{a}$ is integral. If γ in $c\mathfrak{a}$ is such that $(\gamma)/c\mathfrak{a} = (c^{-1}\gamma)/\mathfrak{a}$ is prime to $\mathfrak{b}$ then $\alpha = c^{-1}\gamma$ meets the requirements of Corollary 2; and if δ in $c\mathfrak{a}$ is such that $c\mathfrak{a} = (\delta, c\alpha')$ then $\alpha = c^{-1}\delta$ meets the requirements of Corollary 3.

It follows from Theorem 7 that there is an isomorphism (though not a canonical one)

$$\mathfrak{p}^{n-1}/\mathfrak{p}^n \approx \mathfrak{o}/\mathfrak{p} \tag{5}$$

for any non-zero prime ideal $\mathfrak{p}$ and any integer $n > 0$. For fix β in $\mathfrak{p}^{n-1}$ but not in $\mathfrak{p}^n$; the map $\mathfrak{o} \to \mathfrak{p}^{n-1}/\mathfrak{p}^n$ given by $\alpha \mapsto \alpha\beta$ has kernel $\mathfrak{p}$, so we need only prove that it is onto. Given any γ in $\mathfrak{p}^{n-1}$, by Corollary 1 to Theorem 7 we can find γ_1 satisfying

$$\gamma_1 \equiv \gamma \bmod \mathfrak{p}^n, \quad \gamma_1 \equiv 0 \bmod ((\beta)/\mathfrak{p}^{n-1}).$$

Now $\alpha_1 = \gamma_1/\beta$ is an integer, because any prime power which divides β also divides γ_1; and α_1 maps to γ_1 and hence to $\gamma \bmod \mathfrak{p}^n$.

Now let k be an algebraic number field and $\mathfrak{o}$ the ring of integers in k. It follows from Lemma 1 that any non-zero fractional ideal $\mathfrak{a}$ is a lattice in k. We now define the **absolute norm** Norm $\mathfrak{a}$ of $\mathfrak{a}$; note the capital letter. Let V be the **R**-vector space $k \otimes_{\mathbf{Q}} \mathbf{R}$, on which volume is defined up to multiplication by an arbitrary constant. Both $\mathfrak{o}$ and $\mathfrak{a}$ are lattices in V, so we can write

$$\text{Norm } \mathfrak{a} = \text{vol}(V/\mathfrak{a})/\text{vol}(V/\mathfrak{o}).$$

If $\mathfrak{a}$ is integral, then

$$\text{Norm } \mathfrak{a} = [\mathfrak{o} : \mathfrak{a}]. \tag{6}$$

By convention, and despite this definition, we write $\text{Norm}\,(0) = 0$ where (0) is the zero ideal. If $\alpha \neq 0$ is in k and $\alpha_1, \ldots, \alpha_n$ are a base for $\mathfrak{a}$ as a **Z**-module, then $\alpha\alpha_1, \ldots, \alpha\alpha_n$ are a base for $\alpha\mathfrak{a}$. Using the volume defined on page 3, we deduce

$$\text{Norm}(\alpha\mathfrak{a}) = \left|\prod_\sigma \sigma\alpha\right| \text{Norm } \mathfrak{a} = |\text{norm}_{k/\mathbf{Q}}\alpha|\,\text{Norm } \mathfrak{a} \tag{7}$$

and in particular $\text{Norm}((\alpha)) = |\text{norm}_{k/\mathbf{Q}}\alpha|$. We could instead have used (6) and (7) to define the absolute norm; this is more algebraic but clumsier.

Lemma 8 *If $\mathfrak{a}_1, \mathfrak{a}_2$ are fractional ideals in k, then*

$$(\text{Norm } \mathfrak{a}_1)(\text{Norm } \mathfrak{a}_2) = \text{Norm}(\mathfrak{a}_1\mathfrak{a}_2).$$

Proof In view of (7) it is enough to prove this when the $\mathfrak{a}_\mu$ are integral and non-zero. If $\mathfrak{a} = \prod \mathfrak{p}_\mu^{n_\mu}$ is integral, then by (5), (6) and Theorem 7

$$\text{Norm } \mathfrak{a} = \prod [\mathfrak{o} : \mathfrak{p}_\mu]^{n_\mu} = \prod (\text{Norm } \mathfrak{p}_\mu)^{n_\mu}.$$

The lemma now follows by factorizing $\mathfrak{a}_1$ and $\mathfrak{a}_2$. $\square$

For some applications it is convenient to generalize $\mathfrak{o}_k$, the purpose being to enable us to ignore certain 'bad' primes. If $\mathfrak{p}$ is a prime ideal of $\mathfrak{o}_k$, we say that α in k is **integral** at $\mathfrak{p}$ if $\alpha = \alpha_1/\alpha_2$ where α_1, α_2 are in $\mathfrak{o}_k$ and α_2 is not in $\mathfrak{p}$; provided that $\alpha \neq 0$, this is equivalent to saying that $\mathfrak{p}$ does not occur among the $\mathfrak{q}_\nu$ in the factorization (3) of (α). More generally, let S be any set of primes, and define $\mathfrak{o}_S$ to be the set of elements of k integral at each prime outside S. (This notation must not be confused with the much more important $\mathfrak{o}_\mathfrak{p}$ introduced in Chapter 2.) It is easy to see that $\mathfrak{o}_S$ is Dedekind. The ideals of $\mathfrak{o}_S$ are just the $\mathfrak{a}_S = \mathfrak{a} \otimes_\mathfrak{o} \mathfrak{o}_S$ where $\mathfrak{a}$ is an ideal of $\mathfrak{o}$; for we can recover one possible $\mathfrak{a}$ from any ideal $\mathfrak{b}$ of $\mathfrak{o}_S$ because $\mathfrak{a} = \mathfrak{b} \cap \mathfrak{o}$ implies $\mathfrak{b} = \mathfrak{a}_S$. To any fractional ideal $\mathfrak{a}$ in k there corresponds the fractional ideal $\mathfrak{a}_S = \mathfrak{a} \otimes_\mathfrak{o} \mathfrak{o}_S$; but recovering a possible $\mathfrak{a}$ from a fractional ideal for $\mathfrak{o}_S$ is untidy and is best done by means of (3). The prime ideals of $\mathfrak{o}_S$ are just the $\mathfrak{p}_S$ where $\mathfrak{p}$ is a prime ideal of $\mathfrak{o}$ not in

S; to derive from (2) a factorization of $\mathfrak{a}_S$ we delete from the right hand side any primes in S and put a subscript S on the others.

There is a general theorem that if $\mathfrak{o}$ is a Dedekind domain with quotient field k, and if K is a finite algebraic extension of k and $\mathfrak{O}$ the integral closure of $\mathfrak{o}$ in K, then $\mathfrak{O}$ is Dedekind; taking $\mathfrak{o}, k$ to be $\mathbf{Z}, \mathbf{Q}$ it follows that the ring of integers of any algebraic number field is a Dedekind domain. But the proof of this general theorem is tedious and there is an easy shortcut.

Theorem 8 *If $\mathfrak{o}$ is the ring of integers of an algebraic number field k then $\mathfrak{o}$ is Dedekind.*

Proof We have only to prove that any non-zero prime ideal $\mathfrak{p}$ in $\mathfrak{o}$ is maximal — in other words that $\mathfrak{o}/\mathfrak{p}$ is a field. It is an integral domain, so we only have to prove the existence of inverses. Let α be in $\mathfrak{o}$ but not in $\mathfrak{p}$. Multiplication by α gives a map of $\mathfrak{o}/\mathfrak{p}$ to itself with trivial kernel; since $\mathfrak{o}/\mathfrak{p}$ is finite such a map must be onto, so the class of α has an inverse. ☐

Since $\mathfrak{o}/\mathfrak{p}$ is a finite field, it contains p^f elements for some rational prime p and some $f = f_{\mathfrak{p}} > 0$. Clearly $\mathfrak{p} \supset (p)$, so let $\mathfrak{p}^e$ be the exact power of $\mathfrak{p}$ which divides (p). The primes with $e_{\mathfrak{p}} > 1$ are called **ramified**; we shall see later that for each k there are only finitely many of them. We can factorize (p) as a finite product $(p) = \prod \mathfrak{p}^{e_{\mathfrak{p}}}$; taking Norms on both sides and writing $n = [k : \mathbf{Q}]$ we obtain

$$n = \sum e_{\mathfrak{p}} f_{\mathfrak{p}} \tag{8}$$

where the sum is taken over all $\mathfrak{p}$ dividing a fixed p. The analogue for extensions K/k will be proved as the Corollary to Lemma 13.

Let $\mathfrak{o}$ be a Dedekind domain with quotient field k, and let J_k be the group of non-zero fractional ideals of k. The **ideal class group** of k is

$$\mathcal{C}_k = J_k/(\text{group of non-zero principal ideals}).$$

If k is an algebraic number field and $\mathfrak{o}$ its ring of integers, we shall show below that the order of J_k is finite; it is called the **class number** of k and is denoted by h. There is an exact sequence

$$0 \to \mathfrak{o}^* \to k^* \to J_k \to \mathcal{C}_k \to 0,$$

where $\mathfrak{o}^*$ is the group of units of $\mathfrak{o}$. For a general Dedekind domain, we can say nothing useful about the two outer terms. But if $\mathfrak{o}$ is the ring of integers of an algebraic number field k (or even if $\mathfrak{o} = \mathfrak{o}_S$ for some set S of primes in such a k), they have important properties. We consider units in

§3; the main theorem about $\mathcal{C}_k$ is Theorem 9 below, though it too will be refined in §3.

Let $[k : \mathbf{Q}] = n$ and fix a base $\alpha_1, \ldots, \alpha_n$ for $\mathfrak{o}_k$ as a **Z**-module. Any β in k can be written

$$\beta = b_1\alpha_1 + \cdots + b_n\alpha_n \quad \text{with } b_\nu \text{ in } \mathbf{Q};$$

we define the **height** of β to be $H(\beta) = \sum |b_\nu|$, which gives a metric on k. Write $C = \max H(\alpha_\mu \alpha_\nu)$. If β is in $\mathfrak{o}_k$ then $H(\beta)$ is in **Z**, as is C.

Lemma 9 *Let $\mathfrak{a}$ be a non-zero fractional ideal in an algebraic number field k and α a non-zero element of $\mathfrak{a}$ of minimal height. For any β in $\mathfrak{a}$ we can find γ in $\mathfrak{o}_k$ and m in **Z** such that*

$$H(m\beta/\alpha - \gamma) \leqslant (C+1)^{-1} \quad and \quad 0 < m \leqslant M, \tag{9}$$

where $M = (n(C+1))^n + 1$.

Proof If we multiply $\mathfrak{a}$, α and β by the same integer N, we alter neither the hypotheses nor the conclusions; so we can assume that $\mathfrak{a}$ is integral. In particular, H is integral on $\mathfrak{a}$, so its minimum is attained and α exists. We now use the pigeonhole principle. For any γ_m in $\mathfrak{o}$ write

$$m\beta/\alpha - \gamma_m = c_1^{(m)}\alpha_1 + \cdots + c_n^{(m)}\alpha_n \text{ with } c_\nu^{(m)} \text{ in } \mathbf{Q};$$

given m we can choose γ_m in $\mathfrak{o}$ so that $0 \leqslant c_\nu^{(m)} < 1$ for each ν and thus determine a point $P_m = (c_1^{(m)}, \ldots, c_n^{(m)})$ in the unit cube. Partition this cube into $M - 1$ subcubes of side $(n(C+1))^{-1}$, each defined by

$$r_\nu/(n(C+1)) \leqslant c_\nu < (r_\nu + 1)/(n(C+1))$$

for some integers r_ν with $0 \leqslant r_\nu < n(C+1)$. Two of $P_1, \ldots, P_M$ must lie in the same subcube — say P_{m_1} and P_{m_2} with $m_1 < m_2$. Writing $m = m_2 - m_1$, $\gamma = \gamma_{m_2} - \gamma_{m_1}$ we obtain (9).   □

Theorem 9 *If k is an algebraic number field the order of $\mathcal{C}_k$ is finite.*

Proof For any $\epsilon = \sum e_\mu \alpha_\mu$, $\epsilon' = \sum e'_\nu \alpha_\nu$ in k we have

$$H(\epsilon\epsilon') = H\left(\sum\sum e_\mu e'_\nu \alpha_\mu \alpha_\nu\right)$$
$$\leqslant \sum\sum |e_\mu e'_\nu| H(\alpha_\mu \alpha_\nu) \leqslant CH(\epsilon)H(\epsilon')$$

in the notation above. It now follows from (9) that

$$H(m\beta - \alpha\gamma) \leqslant CH(\alpha)H(m\beta/\alpha - \gamma) \leqslant CH(\alpha)/(C+1).$$

But $m\beta - \alpha\gamma$ is in $\mathfrak{a}$, and must therefore vanish because α has minimal height among the non-zero elements of $\mathfrak{a}$. It follows that $(M!)\beta$ is a multiple of α for every β in $\mathfrak{a}$. Now consider $\mathfrak{a}_1 = (M!)\mathfrak{a}/\alpha$. It is an integral ideal in the class of $\mathfrak{a}$, and it contains $M!$ because $\mathfrak{a}$ contains α; so it is a union of cosets of $(M!)\mathfrak{o}$ in $\mathfrak{o}$. But there are only finitely many ideals composed in this way. $\qquad\qquad\Box$

The corresponding result for $(\mathfrak{o}_k)_S$ as defined on page 15 follows at once. If $\mathfrak{a}$ and $\mathfrak{b}$ are in the same ideal class for $\mathfrak{o}$, so are $\mathfrak{a}_S$ and $\mathfrak{b}_S$ for $\mathfrak{o}_S$; so Theorem 9 for $\mathfrak{o}_S$ follows from Theorem 9 for $\mathfrak{o}$.

We shall obtain a much more realistic way of bounding the order of $\mathcal{C}_k$ in Theorem 10 below, whose proof does not depend on Theorem 9. The real point of the proof above lies in the following exercise.

Exercise Without assuming any of the earlier results in this section, define a relation between non-zero fractional ideals $\mathfrak{a}, \mathfrak{a}'$ in a given algebraic number field k by

$$\mathfrak{a} \sim \mathfrak{a}' \text{ if there exist non-zero } \alpha, \alpha' \text{ in } k \text{ such that } \alpha\mathfrak{a} = \alpha'\mathfrak{a}'.$$

Prove that this is an equivalence relation, and that the argument of Lemma 9 and Theorem 9 shows that there are only finitely many equivalence classes. Now proceed as follows.

- If $\mathfrak{a}\mathfrak{b} \subset \alpha\mathfrak{b}$ for some non-zero $\mathfrak{b}$, then $\mathfrak{a} \subset (\alpha)$.

Let $\beta_1, \ldots, \beta_n$ be a base for $\mathfrak{b}$ and for any α' in $\mathfrak{a}$ write $\lambda = \alpha'/\alpha$. Since $\lambda\mathfrak{b} \subset \mathfrak{b}$ there exist $c_{\mu\nu}$ in $\mathbf{Z}$ such that $\lambda\beta_\mu = \sum c_{\mu\nu}\beta_\nu$; so λ satisfies $\det(\lambda I - C) = 0$ where C is the matrix of the $c_{\mu\nu}$. Hence λ is an integer.

- If $\mathfrak{a}\mathfrak{b} = \alpha\mathfrak{b}$ for some non-zero $\mathfrak{b}$, then $\mathfrak{a} = (\alpha)$.

Write $\mathfrak{a}' = \alpha^{-1}\mathfrak{a}$, which we now know to be integral. Thus $\mathfrak{b} = \mathfrak{a}'\mathfrak{b}$, so that $\beta_\nu = \sum \alpha_{\mu\nu}\beta_\nu$ with $\alpha_{\mu\nu}$ in $\mathfrak{a}'$. Hence $\det(I - A) = 0$ where A is the matrix of the $\alpha_{\mu\nu}$; expanding this determinant shows that 1 is in $\mathfrak{a}'$.

- For any fractional ideal $\mathfrak{a}$, there exists $\mathfrak{b} \neq (0)$ such that $\mathfrak{a}\mathfrak{b}$ is principal.

Among the $\mathfrak{a}^m$ with $m \geqslant 0$ there must be two in the same equivalence class — say $\mathfrak{a}^{m_1}$ and $\mathfrak{a}^{m_2}$ with $m_1 < m_2$. Thus $\alpha_1\mathfrak{a}^{m_1} = \alpha_2\mathfrak{a}^{m_2}$ for some α_1, α_2, whence $(\alpha_1) = \alpha_2\mathfrak{a}^{m_2-m_1}$ and we can take $\mathfrak{b} = \mathfrak{a}^{m_2-m_1-1}$.

Now deduce the main results of this section for the special case when $\mathfrak{o} = \mathfrak{o}_k$, the ring of integers of an algebraic number field k.

3 Embedding in the complex numbers

Let k be an algebraic number field with $[k : \mathbf{Q}] = n$; in this section we exploit systematically the embedding $k \to \mathbf{R}^{r_1} \times \mathbf{C}^{r_2} \approx \mathbf{R}^n$ already introduced on page 3. For convenience we order the embeddings $\sigma_\nu : k \to \mathbf{C}$ so that $\sigma_1, \ldots, \sigma_{r_1}$ are real and that $\sigma_\nu, \sigma_{\nu+r_2}$ are complex conjugate when $r_1 < \nu \leqslant r_1 + r_2$. We shall need the following elementary result.

Lemma 10 *Let Λ be a lattice in $V \approx \mathbf{R}^n$ and let S be a bounded closed convex subset of V symmetric about the origin; then S contains a point of Λ other than the origin provided $\operatorname{vol}(S) \geqslant 2^n \operatorname{vol}(V/\Lambda)$.*

Proof Assume first that $\operatorname{vol}(S) > 2^n \operatorname{vol}(V/\Lambda)$. If the map $\frac{1}{2}S \to V/\Lambda$ is one-one into, then $\operatorname{vol}(\frac{1}{2}S) \leqslant \operatorname{vol}(V/\Lambda)$ which is false. So there are points P_1, P_2 in $\frac{1}{2}S$ with the same image in V/Λ. By symmetry $-P_2$ is in $\frac{1}{2}S$, so that by convexity $\frac{1}{2}(P_1 - P_2)$ is in $\frac{1}{2}S$. Thus $P_1 - P_2$ is in both S and Λ.

If $\operatorname{vol}(S) = 2^n \operatorname{vol}(V/\Lambda)$ the same argument shows that $(1 + \epsilon)S$ contains a point $Q_\epsilon \neq 0$ of Λ for any $\epsilon > 0$. But for $\epsilon < 1$ the candidates for Q_ϵ lie in the bounded discrete set $2S \cap \Lambda$, so they belong to a finite set. Hence there is a point $Q \neq 0$ of Λ which belongs to $(1 + \epsilon)S$ for arbitrarily small ϵ; and since S is closed, Q must belong to S.   □

Theorem 10 *Let $\mathfrak{o}$ be the ring of integers of an algebraic number field k of discriminant d.*

(i) *There is a constant $C = C_{r_1,r_2}$ depending only on r_1, r_2 such that each ideal class of k contains an integral ideal of Norm at most $C|d|^{1/2}$.*

(ii) *The order of $\mathcal{C}_k$ is finite.*

Proof Define a map $k \to V \approx \mathbf{R}^n$ by sending α in k to

$$(\sigma_1\alpha, \ldots, \sigma_{r_1}\alpha, \Re\sigma_{r_1+1}\alpha, \ldots, \Re\sigma_{r_1+r_2}\alpha, \Im\sigma_{r_1+1}\alpha, \ldots, \Im\sigma_{r_1+r_2}\alpha)$$

where the σ_ν are ordered as at the beginning of this section. If $\mathfrak{a}$ is any non-zero fractional ideal in k, it follows from Lemma 1 and the remarks on page 10 that the image of $\mathfrak{a}$ in V is a lattice. Let S be any closed bounded convex subset of V symmetric with respect to the origin; let v be the volume of S and M the maximum of

$$|x_1 \cdots x_{r_1}(x_{r_1+1}^2 + x_{r_1+r_2+1}^2) \cdots (x_{r_1+r_2}^2 + x_n^2)|$$

in S. The volume of $V/\mathfrak{a}^{-1}$ is $c = 2^{-r_2}|d|^{1/2}/\operatorname{Norm}\mathfrak{a}$. For $\lambda = 2(c/v)^{1/n}$ the volume of λS is $2^n c$; so by Lemma 10 there exists $\alpha \neq 0$ in $\mathfrak{a}^{-1} \cap \lambda S$, whence $|\text{norm}\,\alpha| \leqslant \lambda^n M$. But $\alpha\mathfrak{a}$ is an integral ideal in the same class as $\mathfrak{a}$, and

$$\text{Norm}(\alpha\mathfrak{a}) = |\text{norm}\,\alpha|\,\text{Norm}\,\mathfrak{a} \leqslant 2^{r_1+r_2} M v^{-1}|d|^{1/2}.$$

This proves (i), and since there are only finitely many integral ideals in k with Norm less than a preassigned bound, (ii) follows. $\qquad\square$

If one is going to use this theorem to compute h (and in general no other method is known), then it is desirable to make $C_{r_1,r_2} = 2^{r_1+r_2}M/v$ as small as possible. An efficient choice of S is

$$|x_1| + \cdots + |x_{r_1}| + 2\sum_{\nu=r_1+1}^{r_1+r_2} \sqrt{x_\nu^2 + x_{\nu+r_2}^2} \leqslant 1,$$

which gives

$$C_{r_1,r_2} = \left(\frac{4}{\pi}\right)^{r_2} \frac{n!}{n^n}.$$

The details of this calculation are given in the exercise below. Applying (i) to the principal ideal class and noting that $C_{r_1,r_2} < 1$, we obtain

Corollary *For every algebraic number field k, $|d| \geqslant (n^n/n!)(\tfrac{1}{4}\pi)^{r_2} > 1$.*

Thus the discriminant of an algebraic number field k with $[k:\mathbf{Q}] = n$ grows at least exponentially with n. It was shown by Šafarevič that no stronger statement can be true: see [CF], Chapter IX.

Exercise Prove that the S chosen above is convex, and that $M = n^{-n}$. [For convexity, use the triangle inequality and

$$\lambda\sqrt{x_1^2 + y_1^2} + (1-\lambda)\sqrt{x_2^2 + y_2^2}$$
$$\geqslant \sqrt{(\lambda x_1 + (1-\lambda)x_2)^2 + (\lambda y_1 + (1-\lambda)y_2)^2}$$

for $0 \leqslant \lambda \leqslant 1$. In polar coordinates this last inequality is

$$\lambda r_1 + (1-\lambda)r_2 \geqslant \sqrt{\lambda^2 r_1^2 + 2\lambda(1-\lambda)r_1 r_2 \cos(\theta_1 - \theta_2) + (1-\lambda)^2 r_2^2},$$

which is trivial. That $M \leqslant n^{-n}$ follows from the inequality between the arithmetic and geometric means.]

Transforming pairs $x_\nu, x_{\nu+r_2}$ from Cartesian to polar coordinates, show also that $v = 2^{r_1}(2\pi)^{r_2} D_{r_1,r_2}(n)$ where

$$D_{\ell,m}(t) = \int \cdots \int_{\mathcal{R}_{\ell,m}(t)} y_1 \cdots y_m\, dx_1 \cdots dx_\ell\, dy_1 \cdots dy_m$$

and $\mathcal{R}_{\ell,m}(t)$ is given by $x_\rho \geqslant 0\,(1 \leqslant \rho \leqslant \ell)$, $y_\rho \geqslant 0\,(1 \leqslant \rho \leqslant m)$ and

$$x_1 + \cdots + x_\ell + 2(y_1 + \cdots + y_m) \leqslant t.$$

Prove that

$$D_{\ell,m}(t) = \int_0^t D_{\ell-1,m}(t-x)dx = \int_0^{t/2} D_{\ell,m-1}(t-2y)y\,dy$$

and deduce $D_{\ell,m}(t) = 4^{-m}t^{\ell+2m}/(\ell+2m)!$ by induction. $\square$

For small values of n, more is known. For $r_1 = 0, r_2 = 1$ we can take $C_{0,1}^{-1} = \sqrt{3}$ and this appears to be all that can be said. In all other cases, there is known or conjectured to be an isolation theorem. To explain this, it will be convenient to write temporarily

$$C_{\mathfrak{c}} = (\text{least Norm of an integral ideal in the class } \mathfrak{c})/|d_k|^{1/2},$$

where $\mathfrak{c}$ is an ideal class for the field k. For $r_1 = 2, r_2 = 0$ we have $C_{\mathfrak{c}}^{-1} \geqslant \sqrt{5}$, but $C_{\mathfrak{c}}^{-1} \geqslant \sqrt{8}$ except when $\mathfrak{c}$ is the unique ideal class in $\mathbf{Q}(\sqrt{5})$; the least point of accumulation of the $C_{\mathfrak{c}}^{-1}$ is 3, and all the classes $\mathfrak{c}$ with $C_{\mathfrak{c}}^{-1} < 3$ can be specified. (For proofs of some of these statements, see the exercise on page 62.) For $r_1 = 3, r_2 = 0$ we have $C_{\mathfrak{c}}^{-1} \geqslant 9$ except when $\mathfrak{c}$ is the unique ideal class in the field $\mathbf{Q}(2\cos\frac{2\pi}{7})$ with $d = 49$. For $r_1 = r_2 = 1$ we have $C_{\mathfrak{c}}^{-1} \geqslant \sqrt{23}$. But the methods by which these results are established have nothing to do with algebraic number theory. They belong to the Geometry of Numbers, which was once fashionable but which (except for the widely applied Lemma 10) ceased to be so some thirty years ago in England and considerably earlier elsewhere.

All these statements are best possible, in the sense that the constants cannot be improved. It is conjectured that both the cases $r_1 = r_2 = 1$ and $r_1 = 0, r_2 = 2$ have properties like those which have been proved for $r_1 = 2, r_2 = 0$. More spectacularly, it is conjectured that for each pair r_1, r_2 with $r_1 + r_2 > 2$ the $C_{\mathfrak{c}}^{-1}$ have no finite point of accumulation.

A **unit** of $\mathfrak{o}$ is by definition an element $\alpha \neq 0$ of $\mathfrak{o}$ such that α^{-1} is also in $\mathfrak{o}$; thus the units form a group which in standard notation is just $\mathfrak{o}^*$. An alternative and more useful definition is that a unit is any element α of $\mathfrak{o}$ such that $\mathrm{norm}_{k/\mathbf{Q}}(\alpha) = \pm 1$. For if α is a unit then $\mathrm{norm}(\alpha)$ and $\mathrm{norm}(\alpha^{-1})$ are elements of $\mathbf{Z}$ whose product is 1; conversely, if α is in $\mathfrak{o}$ with $\mathrm{norm}(\alpha) = \pm 1$ then $\pm\alpha^{-1}$ is equal to $\mathrm{norm}(\alpha)/\alpha$, which is a product of powers of conjugates of α and therefore an integer.

Theorem 11 *The group of units in* o *is the product of the group of roots of unity in* o, *which is cyclic and finite, and a free group on* $r_1 + r_2 - 1$ *generators.*

Proof The map $o^* \to \mathbf{R}^{r_1+r_2}$ defined by

$$\alpha \mapsto (\log|\sigma_1\alpha|, \ldots, \log|\sigma_{r_1+r_2}\alpha|) \tag{10}$$

is a homomorphism from o^* to the hyperplane

$$Y_1 + \cdots + Y_{r_1} + 2(Y_{r_1+1} + \cdots + Y_{r_1+r_2}) = 0 \tag{11}$$

in $\mathbf{R}^{r_1+r_2}$. Its kernel consists of the α in o^* with $|\sigma\alpha| = 1$ for all σ; so in the standard topology on k it is a bounded subset of the discrete set o and is therefore finite. If its order is N then every element of it is an N-th root of unity; and the kernel is cyclic because the group of N-th roots of unity is cyclic, being generated by $\exp(2\pi i/N)$, and hence so are all its subgroups.

It remains to prove that the image of the map (10) is a lattice in the hyperplane (11). For this, let $\mathcal{N}$ be a bounded neighbourhood of the origin in (11) or even in $\mathbf{R}^{r_1+r_2}$. The points of o^* which map into $\mathcal{N}$ have every $|\sigma\alpha|$ bounded; so in the standard topology they lie in the intersection of o and a bounded set, and hence they form a finite set. Thus the image is discrete. It remains to show that the image of (10) spans (11). For this it is enough to prove the following assertion:-

• Given any real $\lambda_1, \ldots, \lambda_{r_1+r_2}$ not all equal, there is a unit η with

$$\begin{aligned} f(\eta) =&\lambda_1 \log|\sigma_1\eta| + \cdots + \lambda_{r_1} \log|\sigma_{r_1}\eta| \\ &+ 2(\lambda_{r_1+1} \log|\sigma_{r_1+1}\eta| + \cdots + \lambda_{r_1+r_2} \log|\sigma_{r_1+r_2}\eta|) \neq 0. \end{aligned}$$

Let $\rho_1, \ldots, \rho_{r_1+r_2}$ be positive real numbers such that

$$\rho_1 \cdots \rho_{r_1}(\rho_{r_1+1} \cdots \rho_{r_1+r_2})^2 = (\tfrac{2}{\pi})^{r_2}|d|^{1/2} = A, \tag{12}$$

say, where d is the discriminant of k. In the coordinates defined on page 18, the set S given by

$$|X_\mu| \leqslant \rho_\mu \text{ for } 1 \leqslant \mu \leqslant r_1, \quad |X_\mu^2 + X_{\mu+r_2}^2| \leqslant \rho_\mu^2 \text{ for } r_1 < \mu \leqslant r_1 + r_2$$

is bounded, closed, convex and symmetric with respect to the origin; and its volume is $2^{r_1+r_2}|d|^{1/2}$. By Lemma 10 there is a non-zero integer α in k such that

$$|\sigma_\mu\alpha| \leqslant \rho_\mu \text{ for } 1 \leqslant \mu \leqslant r_1 + r_2;$$

and $|\mathrm{norm}_{k/\mathbf{Q}}\alpha| \leqslant A$ now follows from (12). Since also $|\mathrm{norm}_{k/\mathbf{Q}}\alpha| \geqslant 1$,

$$|X_\mu| \geqslant A^{-1}\rho_\mu \text{ for } 1 \leqslant \mu \leqslant r_1,$$
$$|X_\mu^2 + X_{\mu+r_2}^2| \geqslant A^{-1}\rho_\mu^2 \text{ for } r_1 < \mu \leqslant r_1 + r_2.$$

Let us say temporarily that α_1, α_2 in $\mathfrak{o}$ are equivalent if α_1/α_2 is a unit. The elements in an equivalence class are those which generate a particular principal ideal, and up to sign their norm is the Norm of that principal ideal; so there are only finitely many equivalence classes whose Norms are bounded by A, because each of them corresponds to a subgroup of $\mathfrak{o}$ of index at most A. Let $\beta_1, \ldots, \beta_N$ be representatives of these classes. The α generated above lies in one of these classes, so $\eta = \alpha/\beta_\nu$ is a unit for some ν. But now $f(\eta) = f(\alpha) - f(\beta_\nu)$ and this differs from

$$\lambda_1 \log\rho_1 + \cdots + \lambda_{r_1}\log\rho_{r_1} + 2(\lambda_{r_1+1}\log\rho_{r_1+1} + \cdots + \lambda_{r_1+r_2}\log\rho_{r_1+r_2})$$

by at most $B = |f(\beta_\nu)| + (\log A)\sum|\lambda_\mu|$, which does not depend on the ρ_μ. We can choose the ρ_μ so that (12) holds and the last displayed expression exceeds B in absolute value; this ensures $f(\eta) \neq 0$. $\qquad\square$

As it stands, this argument is not constructive because we have no way of writing down a complete set of representatives $\beta_1, \ldots, \beta_N$. But by a slight modification of the argument, we can find $r_1 + r_2 - 1$ independent units; for by choosing different sets of ρ_μ we can find as many α with $|\mathrm{norm}_{k/\mathbf{Q}}\alpha| \leqslant A$ as we wish, and suitable quotients of these will be units. Once we have $r_1 + r_2 - 1$ independent units, it is in principle straightforward to find a base for the group of units; for we can quantify the statement that if α is a unit and the image of α under the map (10) is near enough to the origin then α is a root of unity. But in practice such calculations can be quite tedious.

Corollary *Let S be a finite set, consisting of m primes in $\mathfrak{o}$. Then the group of units in $\mathfrak{o}_S$ is the product of the group of roots of unity in $\mathfrak{o}_S$ (which is the same as for $\mathfrak{o}$) and a free group on $r_1 + r_2 + m - 1$ generators.*

Proof Let $\mathfrak{p}_1, \ldots, \mathfrak{p}_m$ be the elements of S. There is an exact sequence

$$0 \to \mathfrak{o}^* \to \mathfrak{o}_S^* \to \{\text{free group generated by the } \mathfrak{p}_\mu\};$$

and the image of the right hand map contains each $\mathfrak{p}_\mu^h$ and is therefore free of rank m. $\qquad\square$

Since the units give rise to a lattice Λ in the vector space V defined by (11) and there is a canonical measure on V, we expect the volume of V/Λ

to be of interest. This volume is called the **regulator** R of k. To write it down explicitly, choose any base $\eta_1, \ldots, \eta_{r_1+r_2-1}$ for the group

$$\{\text{units}\}/\{\text{roots of unity}\}$$

and form the $(r_1 + r_2 - 1) \times (r_1 + r_2)$ matrix whose ν-th row is

$$(\log|\sigma_1\eta_\nu|, \ldots, \log|\sigma_{r_1}\eta_\nu|, 2\log|\sigma_{r_1+1}\eta_\nu|, \ldots, 2\log|\sigma_{r_1+r_2}\eta_\nu|).$$

Now delete any column; it does not matter which, since the sum of the columns is 0. The regulator R is then defined to be the absolute value of the determinant of the resulting matrix.

4 Change of fields

Let $k \subset K$ be algebraic number fields; it is natural to ask how far the concepts which we have introduced for extensions $k/\mathbf{Q}$ can be extended to K/k. For this, one needs to be able to lift certain objects from k to K. Elements of k are automatically elements of K, and the property of being integral does not depend on the field we are working in; but there is more difficulty with ideals. There is only one sensible way of lifting a fractional or integral ideal $\mathfrak{a}$ of k to K; this is by means of the **conorm**. We define $\mathfrak{A} = \text{conorm}_{K/k}\mathfrak{a}$ to be the smallest (fractional or integral) ideal of K which contains all the elements of $\mathfrak{a}$; equivalently we could define it as $\mathfrak{a} \otimes_{\mathfrak{o}} \mathfrak{D}$. It is trivial to check that taking conorms commutes with multiplication of ideals, that if α is in k then $\text{conorm}_{K/k}(\alpha) = (\alpha)$ where the (α) on the left is in k and the (α) on the right is in K, and that the tower law

$$\text{if } K \supset L \supset k \text{ then } \text{conorm}_{K/L}(\text{conorm}_{L/k}\mathfrak{a}) = \text{conorm}_{K/k}\mathfrak{a}$$

holds for $\mathfrak{a}$ in k. In other words, the notation is reasonably foolproof. By abuse of language, one often writes $\mathfrak{a}$ where one should write $\text{conorm}_{K/k}\mathfrak{a}$. In consequence, we sometimes need to write $\text{Norm}_k\mathfrak{a}$ instead of $\text{Norm}\,\mathfrak{a}$ to make it clear which field we regard $\mathfrak{a}$ as belonging to.

This process gives an apparently attractive way of getting rid of ideals and working purely with elements of $\bar{k}$:

Theorem 12 *Given any algebraic number field k there is a finite extension $K \supset k$ such that for every ideal $\mathfrak{a}$ in k, $\text{conorm}_{K/k}\mathfrak{a}$ is principal.*

Proof Let $\mathfrak{a}_1, \mathfrak{a}_2$ be ideals of k in the same ideal class, so that there are non-zero β_1, β_2 in k such that $(\beta_1)\mathfrak{a}_1 = (\beta_2)\mathfrak{a}_2$; then

$$(\beta_1)\text{conorm}_{K/k}\mathfrak{a}_1 = (\beta_2)\text{conorm}_{K/k}\mathfrak{a}_2$$

for any $K \supset k$. Thus if one ideal in an ideal class of k becomes principal in K, they all do. Now let $\mathfrak{a}_1, \ldots, \mathfrak{a}_h$ be representatives of the ideal classes in k; the $\mathfrak{a}_i^h$ are all principal in k, so $\mathfrak{a}_i^h = (\alpha_i)$ for some α_i in k. Let

$$K = k(\sqrt[h]{\alpha_1}, \ldots, \sqrt[h]{\alpha_h});$$

then $(\mathrm{conorm}_{K/k}\mathfrak{a}_i)^h = (\alpha_i)$ and so $\mathrm{conorm}_{K/k}\mathfrak{a}_i = (\sqrt[h]{\alpha_i})$ in K. $\square$

Unfortunately, this technique for describing divisibility is so clumsy as to be unusable. But the theorem has some historical interest, as being one of the first signposts towards class field theory.

If we know that an ideal $\mathfrak{A}$ in K has the form $\mathrm{conorm}_{K/k}\mathfrak{a}$ then we can recover $\mathfrak{a}$ as follows.

Lemma 11 *Let $K \supset k$ and let $\mathfrak{a}$ be any fractional ideal in k. Then*

$$(\mathrm{conorm}_{K/k}\mathfrak{a}) \cap k = \mathfrak{a}.$$

Proof Since $\mathrm{conorm}_{K/k}\mathfrak{a} \supset \mathfrak{a}$ set-theoretically, all we have to prove is that if $\beta = \sum \alpha_\mu A_\mu$ is in k, where the α_μ are in $\mathfrak{a}$ and the A_μ are integers in K, then β is in $\mathfrak{a}$. Extending K if necessary, we can assume K normal over k. Let $\sigma_1, \ldots, \sigma_n$ be the elements of $\mathrm{Gal}(K/k)$; then

$$\beta^n = \prod_\nu (\sigma_\nu \beta) = \prod_\nu \left(\sum_\mu \alpha_\mu \sigma_\nu A_\mu \right).$$

On the right hand side, the coefficient of any monomial in the α_μ is an algebraic integer in K invariant under $\mathrm{Gal}(K/k)$, and hence an integer in k. Thus β^n is in $\mathfrak{a}^n$, and so β is in $\mathfrak{a}$. $\square$

If $\mathfrak{A}$ is an ideal in K, how should we define $\mathrm{norm}_{K/k}\mathfrak{A}$? If possible we should form the product of $\mathfrak{A}$ and all its conjugates, and transfer the result back to k. To do this, we should choose an extension L of K normal over k, denote the distinct k-homomorphisms $K \to L$ by $\sigma_1, \ldots, \sigma_n$, form $\prod \mathrm{conorm}_{L/\sigma_\nu K}(\sigma_\nu \mathfrak{A})$, which is an ideal in L, and show by some analogue of the Fundamental Theorem of Galois Theory that this is $\mathrm{conorm}_{L/k}\mathfrak{a}$ for some ideal $\mathfrak{a}$ in k. But in this simple form the last step does not work; for in contrast with what happens with elements of L, an ideal of L which is invariant under every element of $\mathrm{Gal}(L/k)$ need not be the conorm of an ideal in k. Suppose for example that $k = \mathbf{Q}$ and $L = \mathbf{Q}(\sqrt{-5})$, so that $\mathfrak{O}_L = \mathbf{Z}[\sqrt{-5}]$. The ideal $\mathfrak{A} = (1+\sqrt{-5}, 1-\sqrt{-5})$ consists of all $m+n\sqrt{-5}$ with $m+n$ even; so it is invariant under $\mathrm{Gal}(L/\mathbf{Q})$ and $\mathfrak{A} \cap \mathbf{Q} = (2)$. But $\mathfrak{A} \neq \mathrm{conorm}_{L/\mathbf{Q}}(2)$, so $\mathfrak{A}$ is not a conorm for $L/\mathbf{Q}$. In fact $\mathfrak{A}^2 = (2)$ in L.

The way round this difficulty depends on the following result:

Lemma 12 *Let* $\mathfrak{a}$ *be the ideal in* k *generated by the* $\mathrm{norm}_{K/k}A$ *where* A *runs through the elements of an ideal* $\mathfrak{A}$ *in* K, *and let* $L \supset K$ *be a normal extension of* k. *Then in the notation above*

$$\mathrm{conorm}_{L/k}\mathfrak{a} = \prod_{\sigma_i} \mathrm{conorm}_{L/\sigma_i K}(\sigma_i \mathfrak{A}). \tag{13}$$

Proof Since $\sigma_i A$ is in $\mathrm{conorm}_{L/\sigma_i K}(\sigma_i \mathfrak{A})$, $\mathrm{norm}_{K/k}A = \prod(\sigma_i A)$ is in the right hand side of (13). Hence there is an integral ideal $\mathfrak{L}$ in L such that

$$\mathrm{conorm}_{L/k}\mathfrak{a} = \mathfrak{L} \prod \mathrm{conorm}_{L/\sigma_i K}(\sigma_i \mathfrak{A}).$$

We must prove that $\mathfrak{L} = (1)$. Let $\mathfrak{B}$ be an ideal in K prime to $\mathfrak{L} \cap K$ and such that $\mathfrak{A}\mathfrak{B}$ is a principal ideal, say $\mathfrak{A}\mathfrak{B} = (B)$ with B in K. Since $\mathfrak{L}$ is fixed under $\mathrm{Gal}(L/k)$, $\mathrm{conorm}_{L/\sigma_i K}(\sigma_i \mathfrak{B})$ is prime to $\mathfrak{L}$ for every σ_i. Since B is in $\mathfrak{A}$,

$$\mathrm{conorm}_{L/k}\mathfrak{a} \supset \left(\prod \sigma_i B\right)$$
$$= \left(\prod \mathrm{conorm}_{L/\sigma_i K}(\sigma_i \mathfrak{B})\right) \left(\prod \mathrm{conorm}_{L/\sigma_i K}(\sigma_i \mathfrak{A})\right),$$

so $\prod \mathrm{conorm}_{L/\sigma_i K}(\sigma_i \mathfrak{B})$ is a multiple of $\mathfrak{L}$. But $\prod \mathrm{conorm}_{L/\sigma_i K}(\sigma_i \mathfrak{B})$ is prime to $\mathfrak{L}$. Hence $\mathfrak{L} = (1)$. $\qquad \square$

In view of this result, we define $\mathrm{norm}_{K/k}\mathfrak{A}$ to be the ideal in k generated by all the $\mathrm{norm}_{K/k}A$ for A in $\mathfrak{A}$. We list the standard properties of norms of ideals, which are analogous to those of norms of numbers:

Lemma 13 *Let* $K \supset L \supset k$ *be algebraic number fields, let* $\mathfrak{A}, \mathfrak{B}$ *be ideals in* K *and* $\mathfrak{a}$ *an ideal in* k, *and let* A *be an element of* K. *Then*

$$\mathrm{norm}_{K/k}(A) = (\mathrm{norm}_{K/k}A), \tag{14}$$
$$\mathrm{norm}_{L/k}(\mathrm{norm}_{K/L}\mathfrak{A}) = \mathrm{norm}_{K/k}\mathfrak{A}, \tag{15}$$
$$\mathrm{norm}_{K/k}(\mathfrak{A}\mathfrak{B}) = (\mathrm{norm}_{K/k}\mathfrak{A})(\mathrm{norm}_{K/k}\mathfrak{B}), \tag{16}$$
$$\mathfrak{a}^{[K:k]} = \mathrm{norm}_{K/k}(\mathrm{conorm}_{K/k}\mathfrak{a}), \tag{17}$$
$$(\mathrm{Norm}_k\mathfrak{a}) = \mathrm{norm}_{k/\mathbb{Q}}\mathfrak{a}. \tag{18}$$

Proof We can assume that $\mathfrak{A}, \mathfrak{B}, \mathfrak{a}, A$ are non-zero, for otherwise these equalities are trivial. Now (14) follows immediately from the definition. In each of the other equations, the right hand side is contained in the left hand side in view of the corresponding equalities for numbers. Now choose ideals $\mathfrak{A}', \mathfrak{B}'$ in K and $\mathfrak{a}'$ in k so that $\mathfrak{A}\mathfrak{A}', \mathfrak{B}\mathfrak{B}', \mathfrak{a}\mathfrak{a}'$ are principal, and multiply the inequalities for undashed letters by the corresponding inequalities for

dashed ones. We obtain statements about numbers, in which equality holds
by (14); so equality must also hold in the component inequalities.   □

Corollary *Let $K \supset k$ be algebraic number fields and let $\mathfrak{p}$ be a prime in k
whose factorization in K is* $\mathrm{conorm}_{K/k}\mathfrak{p} = \prod \mathfrak{P}_i^{e_i}$. *Denote by f_i the degree
of $\mathfrak{O}/\mathfrak{P}_i$ over $\mathfrak{o}/\mathfrak{p}$. Then* $\mathrm{norm}_{K/k}\mathfrak{P}_i = \mathfrak{p}^{f_i}$ *and* $[K:k] = \sum e_i f_i$.

Proof Taking Norms and using the last two results in the lemma,

$$(\mathrm{Norm}_K(\mathrm{conorm}_{K/k}\mathfrak{p})) = \mathrm{norm}_{K/\mathbf{Q}}(\mathrm{conorm}_{K/k}\mathfrak{p})$$
$$= \mathrm{norm}_{k/\mathbf{Q}}\mathfrak{p}^{[K:k]} = (\mathrm{Norm}_k\mathfrak{p})^{[K:k]}$$

and

$$\left(\mathrm{Norm}_K \prod \mathfrak{P}_i^{e_i}\right) = \prod(\mathrm{Norm}_K\mathfrak{P}_i)^{e_i} = \prod(\mathrm{Norm}_k\mathfrak{p})^{e_i f_i}.$$

Comparing these results gives the second assertion. Hence $\mathrm{norm}_{K/k}\mathfrak{P}_i$ is a
power of $\mathfrak{p}$ by (17), and $\mathrm{norm}_{k/\mathbf{Q}}(\mathrm{norm}_{K/k}\mathfrak{P}_i) = \mathrm{norm}_{k/\mathbf{Q}}\mathfrak{p}^{f_i}$ by (15), (18)
and $\mathrm{Norm}_K\mathfrak{P}_i = (\mathrm{Norm}_k\mathfrak{p})^{f_i}$; this gives the first assertion.   □

By analogy, we can define the **relative discriminant** of K over k as
follows. Let $[K:k] = n$ and let $\alpha_1,\dots,\alpha_n$ be n elements of $\mathfrak{O}_K$ linearly
independent over k. Imitating (1), we write

$$\Delta_{K/k}^2(\alpha_1,\dots,\alpha_n) = \det(\mathrm{Tr}_{K/k}(\alpha_\mu \alpha_\nu));$$

then the relative discriminant of $\mathfrak{O}_K$ over $\mathfrak{o}$ is the ideal in k generated
by all the $\Delta_{K/k}^2(\alpha_1,\dots,\alpha_n)$. But this is less interesting than the relative
different introduced in §8.

5 Normal extensions

Throughout this section K, k will be algebraic number fields with K normal
over k, so that $G = \mathrm{Gal}(K/k)$ acts on K and everything derived from it. Let
$\mathfrak{p}$ be a prime ideal in k, and let $\mathrm{conorm}_{K/k}\mathfrak{p} = \prod \mathfrak{P}_i^{e_i}$ be its factorization in
K and f_i the degree of the field $\mathfrak{O}_K/\mathfrak{P}_i$ over $\mathfrak{o}_k/\mathfrak{p}$. In an obvious notation
we have the tower laws $e_{K/\mathbf{Q}} = e_{K/k}e_{k/\mathbf{Q}}$ and $f_{K/\mathbf{Q}} = f_{K/k}f_{k/\mathbf{Q}}$. If σ is in
G then also $\mathfrak{p} = \prod(\sigma\mathfrak{P}_i)^{e_i}$. Since each $\sigma\mathfrak{P}_i$ is a prime ideal in K, the $\sigma\mathfrak{P}_i$
must be a permutation of the $\mathfrak{P}_i$. The key to the results of this section is
as follows.

Theorem 13 *Suppose K is normal over k with $G = \mathrm{Gal}(K/k)$. Then G
acts transitively on the $\mathfrak{P}_i$, all the e_i have the same value e, all the f_i have*

the same value f, *and* $[G] = [K : k] = efg$ *where* g *is the number of distinct prime factors of* $\mathfrak{p}$ *in* K.

Proof Let $\mathfrak{P}$ be one of the $\mathfrak{P}_i$; then $\mathfrak{P} \cap k$ is an ideal in $\mathfrak{o}_k$ containing $\mathfrak{p}$ but not containing 1; it must therefore be equal to $\mathfrak{p}$ because $\mathfrak{p}$ is maximal. By Corollary 1 to Theorem 7, we can choose α in $\mathfrak{P}$ so that $(\alpha)/\mathfrak{P}$ is prime to $\mathfrak{p}$; thus $\text{norm}_{K/k}\alpha$ lies in $\mathfrak{P} \cap k = \mathfrak{p}$, so that each $\mathfrak{P}_i$ divides some $\sigma\alpha$. But the ideal $(\sigma\alpha)$ is the product of $\sigma\mathfrak{P}$ and an ideal prime to $\mathfrak{p}$; hence $\mathfrak{P}_i = \sigma\mathfrak{P}$. Since $[K : k] = efg$ is now a special case of the Corollary to Lemma 13, all the other claims follow at once. $\qquad\square$

In what follows, we fix our attention on one prime factor $\mathfrak{P}$ of $\mathfrak{p}$. Let Z denote the subgroup of G consisting of those σ which fix $\mathfrak{P}$; Z is called the **splitting group** (Zerlegungsgruppe) of $\mathfrak{P}$ and its fixed field K_Z the **splitting field**. The reason for this name comes from

Lemma 14 *Let* H *be a subgroup of* G *and let* K_H *be its fixed field. Then* $\mathfrak{P}_i$ *and* $\mathfrak{P}_j$ *divide the same prime ideal in* K_H *if and only if* $\mathfrak{P}_i = \sigma\mathfrak{P}_j$ *for some* σ *in* H.

Proof 'If' is clearly trivial; for 'only if' we apply the transitivity property of Theorem 13 to the normal extension K/K_H with Galois group H. $\qquad\square$

It follows that K_Z is the smallest field L between k and K such that the prime $\mathfrak{P} \cap L$ in L does not split in K, though it may ramify there. Moreover if e, f, g refer to $\mathfrak{P}$ or a prime in a smaller field divisible by it, then $g(K/K_Z) = 1$ and $[K_Z : k] = [G : Z] = g(K/k)$, so that

$$e(K/K_Z)f(K/K_Z) = [K : K_Z] = e(K/k)f(K/k).$$

By the tower laws $e(K/K_Z) \leqslant e(K/k)$ and $f(K/K_Z) \cdot \leqslant f(K/k)$; so we have equality in both relations. Thus in going from k to K_Z we split off the prime $\mathfrak{P} \cap K_Z$ but do not ramify it or extend its residue field.

Now consider the residue field $\mathfrak{O}_K/\mathfrak{P}$. Every element of the splitting group Z induces an automorphism of $\mathfrak{O}_K/\mathfrak{P}$ which leaves $\mathfrak{o}/\mathfrak{p}$ elementwise fixed. Moreover, if α is in K then the characteristic polynomial of α over K_Z is $\psi(X) = \prod_{\sigma \text{ in } Z}(X - \sigma\alpha)$. Denote reduction mod $\mathfrak{P}$ by a tilde; since we have just shown that $\tilde{\psi}(X)$ is defined over $\mathfrak{o}/\mathfrak{p}$ and all its roots are $\widetilde{\sigma\alpha}$ with σ in Z, all the conjugates of $\tilde{\alpha}$ over $\mathfrak{o}/\mathfrak{p}$ have this form. Choosing α so that $\tilde{\alpha}$ generates $\mathfrak{O}_K/\mathfrak{P}$ we deduce that every automorphism of $\mathfrak{O}_K/\mathfrak{P}$ over $\mathfrak{o}/\mathfrak{p}$ is induced by some σ in Z. Let T be the kernel of the epimorphism

$$Z \to \text{Gal}((\mathfrak{O}_K/\mathfrak{P})/(\mathfrak{o}/\mathfrak{p})),$$

where we know from the Galois theory of finite fields that the right hand side is a cyclic group of order f; T is called the **inertia group** (Trägheitsgruppe) and its fixed field K_T the **inertia field** of $\mathfrak{P}_i$. This proves all but the last clause of

Theorem 14 T *is a normal subgroup of Z, of order e, and Z/T is cyclic of order f. T consists of those elements of Z which induce the identity on $\mathfrak{O}_K/\mathfrak{P}$; these are just the elements of G for which $\alpha \equiv \sigma\alpha \bmod \mathfrak{P}$ for all α in $\mathfrak{O}_K$.*

For the last clause, we need only note that if σ is not in Z then we can choose α so that α is in $\mathfrak{P}$ but not in $\sigma^{-1}\mathfrak{P}$, whence $\alpha \not\equiv \sigma\alpha \bmod \mathfrak{P}$. Now $\mathrm{Gal}(K/K_T) = T$, which is also the inertia group of $\mathfrak{P}$ for the extension K/K_T; thus $f(K/K_T) = 1$, whence $e(K/K_T) = e$ and so $\mathfrak{P} \cap K_T = \mathfrak{P}^e$. Thus $\mathfrak{P} \cap K_Z$ remains prime in K_T, which is why K_T is called the inertia field, but going from K_Z to K_T induces an extension of degree f of the residue field.

We can go further. Choose Π in $\mathfrak{O}_K$ so that $\mathfrak{P}\|\Pi$ and consider those σ in T for which $\sigma\Pi \equiv \Pi \bmod \mathfrak{P}^2$. In view of the last clause of Theorem 14 this property does not depend on the choice of Π, and hence such σ form a group V; it is called the **ramification group** (Verzweigungsgruppe).

Theorem 15 V *is normal in Z, and is the unique Sylow p-subgroup of T; moreover T/V is cyclic and its order divides $\mathrm{Norm}(\mathfrak{P}) - 1$.*

Proof If p is the rational prime underlying $\mathfrak{P}$ then any element of V has order a power of p. For if σ is an element of V other than the identity then we can choose Π so that $\sigma\Pi \neq \Pi$. Thus $\sigma\Pi \equiv \Pi + \alpha\Pi^m \bmod \mathfrak{P}^{m+1}$ for some $m > 1$ and some α in $\mathfrak{O}_K$ not divisible by $\mathfrak{P}$. Iterating, we obtain $\sigma^r\Pi \equiv \Pi + r\alpha\Pi^m \bmod \mathfrak{P}^{m+1}$. Thus σ cannot have order prime to p, and the same happens for any power of σ other than the identity. Now let σ be any element of T and write $\beta_\sigma = \sigma\Pi/\Pi$; then $\widetilde{\beta_\sigma}$ does not depend on the choice of Π and $\sigma \to \widetilde{\beta_\sigma}$ is a homomorphism from T into the cyclic group $(\mathfrak{O}_K/\mathfrak{P})^*$ with kernel V. This implies in particular that V is normal in T. But since $(\mathfrak{O}_K/\mathfrak{P})^*$ has order prime to p, V is p-Sylow. Uniqueness follows from the facts that all Sylow p-subgroups of an arbitrary finite group are conjugate and V is normal in T. That V is normal in Z follows from the facts that T is normal in Z and V is uniquely determined by T. □

It is possible to investigate the structure of V further, by defining the higher ramification groups. For these, and their connection with the different (introduced in §8), see Chapter IV of [Se]. It turns out that a knowledge of the chain of ramification groups associated with $\mathfrak{P}$ tells one the exact power of $\mathfrak{P}$ which divides the different. This both strengthens Theorem 21 in Chapter 2 and proves the assertion made in §8 that the different is a measure of the badness of bad primes. (The assertion is still valid for non-normal extensions, but the proof requires more complicated notation.) Of course, once one has obtained Z the rest of this section is really local theory and might be better expressed in the language of Chapter 2; for in that language $K \cap k_{\mathfrak{p}} = K_Z$ within $K_{\mathfrak{P}}$ and hence Z is canonically isomorphic to $\mathrm{Gal}(K_{\mathfrak{P}}/k_{\mathfrak{p}})$.

If $e = 1$ then T is trivial; and $Z = \mathrm{Gal}((\mathfrak{O}_K/\mathfrak{P})/(\mathfrak{o}/\mathfrak{p}))$, which is now cyclic of order f, has a natural generator given by $\tilde{\alpha} \mapsto \tilde{\alpha}^{\mathrm{Norm}\,\mathfrak{p}}$ for all $\tilde{\alpha}$. The corresponding element of G is called the **Frobenius element** and is denoted by $[\frac{K/k}{\mathfrak{P}}]$. It is uniquely determined as an element of G by

$$\left[\frac{K/k}{\mathfrak{P}} \right] \alpha \equiv \alpha^{\mathrm{Norm}\,\mathfrak{p}} \bmod \mathfrak{P} \tag{19}$$

for all α in $\mathfrak{O}_K$. It obviously has the property

$$\left[\frac{K/k}{\sigma\mathfrak{P}} \right] = \sigma \left[\frac{K/k}{\mathfrak{P}} \right] \sigma^{-1}$$

for every σ in G; thus it is defined up to conjugacy by a knowledge of $\mathfrak{p}$. In particular, if G is abelian then this symbol depends only on $\mathfrak{p}$; in this case it is called the **Artin element** or Artin symbol and is denoted by $(\frac{K/k}{\mathfrak{p}})$. (Note the change from square to round brackets.) By multiplication we can now define the Artin symbol $(\frac{K/k}{\mathfrak{a}})$ for any fractional ideal $\mathfrak{a}$ which involves no ramified prime; and by construction the map $\mathfrak{a} \mapsto (\frac{K/k}{\mathfrak{a}})$ is a homomorphism. The Artin symbol plays a central role in class field theory.

In some contexts one needs to extend the definition of the Frobenius and Artin elements to the case of primes ramified in K/k. There is only one sensible way to do this; that is to define $[\frac{K/k}{\mathfrak{P}}]$ to be the set of all elements of G which satisfy (19). This identifies the Frobenius element as a member of Z/T, or equivalently as a left coset of T in Z or even in G.

It is natural to ask how all these objects behave under change of field. For this purpose we suppose that $K \supset L \supset k$, that K is normal over k with $G = \mathrm{Gal}(K/k)$, that $H = \mathrm{Gal}(K/L)$ and that $\mathfrak{P}$ is a prime ideal of K and $\mathfrak{Q}, \mathfrak{p}$ are the corresponding prime ideals of L, k. If we consider L/k we shall also need to assume that H is normal in G, so that $\mathrm{Gal}(L/k) = G/H$.

Theorem 16 *With the notation above, $Z_{K/L} = Z_{K/k} \cap H$ and similarly for T and V; and $[\frac{K/L}{\mathfrak{P}}] = [\frac{K/k}{\mathfrak{P}}]^{f'}$ where $\mathrm{Norm}\,\mathfrak{Q} = (\mathrm{Norm}\,\mathfrak{p})^{f'}$, so that f'*

is associated with Ω as a factor of $\mathfrak{p}$. If H is normal in G then $Z_{L/k}$ is the image of $Z_{K/k}$ in G/H and similarly for T and V; and $[\frac{L/k}{\Omega}]$ is the image of $[\frac{K/k}{\mathfrak{P}}]$.

Proof The first half of the first sentence is immediate, and the second follows from (19) because $[\frac{K/k}{\mathfrak{p}}]f'$ is the only element of G with the requisite property. Now assume that H is normal in G, and let σ be an element of G. The image of σ in G/H is in $Z_{L/k}$ if and only if $\sigma\mathfrak{P}$ divides Ω; and since H is transitive on the prime ideals in K which divide Ω, this happens if and only if σH meets $Z_{K/k}$. A similar argument works for T, and the result for V follows from the fact that V is the unique Sylow p-subgroup of T. The result for the Frobenius symbol follows trivially from (19). $\qquad\square$

Suppose that we consider another prime factor $\sigma\mathfrak{P}$ of $\mathfrak{p}$ instead of $\mathfrak{P}$; then the Z, T, V for $\sigma\mathfrak{P}$ are obtained from those for $\mathfrak{P}$ by conjugation by σ. In an obvious notation, $Z_{\sigma\mathfrak{P}} = \sigma Z_{\mathfrak{P}}\sigma^{-1}$ and so on. We have already noted that the corresponding statements hold for the Frobenius and Artin elements.

2

Valuations

6 Valuations and completions

For any field k, an **absolute value**, also called a **multiplicative valuation**, will be a map $k \to \mathbf{R}$ denoted by $\alpha \mapsto \|\alpha\|$ which for some $a > 0$ satisfies the conditions $\|\alpha\| > 0$ for $\alpha \neq 0$, $\|0\| = 0$ and

$$\|\alpha\| \cdot \|\beta\| = \|\alpha\beta\|, \quad \|\alpha + \beta\|^a \leqslant \|\alpha\|^a + \|\beta\|^a. \tag{20}$$

Readers are warned that some authors use a more restrictive definition. Two valuations $\|.\|_1$ and $\|.\|_2$ are called **equivalent** if $\|\alpha\|_2 = \|\alpha\|_1^c$ for some fixed $c > 0$ and all α. An equivalence class of valuations is called a **place**. Any valuation makes k into a metric space with metric given by $d(x_1, x_2) = \|x_1 - x_2\|^a$; this metric depends on a, but the induced topology only depends on the place.

Lemma 15 *Distinct places induce distinct topologies on k.*

Proof Suppose that $\|.\|_1$ and $\|.\|_2$ induce the same topology on k. Since

$$\|x\| < 1 \iff \|x^n\| \to 0 \text{ as } n \to \infty \iff x^n \to 0 \text{ as } n \to \infty,$$

$\|x\|_1 < 1$ if and only if $\|x\|_2 < 1$. We can assume that there exists $x_0 \neq 0$ with this property, for otherwise $\|x\|_1 = \|x\|_2 = 1$ for all $x \neq 0$. Define $c > 0$ by $\|x_0\|_2 = \|x_0\|_1^c$ and for any $x \neq 0$ in k with $\|x\|_1 < 1$ let λ satisfy $\|x\|_1 = \|x_0\|_1^\lambda$. If $m/n > \lambda$ then $\|x_0^m/x^n\|_1 < 1$ whence $\|x_0^m/x^n\|_2 < 1$ and $\|x\|_2 > \|x_0\|_2^{m/n}$; similarly if $m/n < \lambda$ then $\|x\|_2 < \|x_0\|_2^{m/n}$. Thus $\|x\|_2 = \|x_0\|_2^\lambda = \|x_0\|_1^{c\lambda} = \|x\|_1^c$. □

At this stage it must appear more natural to get rid of a and replace the second relation (20) by the usual triangle inequality

$$\|\alpha + \beta\| \leqslant \|\alpha\| + \|\beta\|. \tag{21}$$

However on page 35 we shall see that in each place there is a natural distinguished element (called a normalized valuation), and unfortunately it does not always satisfy (21).

In what follows we shall ignore the trivial valuation $\|\alpha\| = 1$ for all $\alpha \neq 0$, which corresponds to the discrete topology. A multiplicative valuation on a field k is called **Archimedean** if k has characteristic 0 and $\|m\| > 1$ for some m in $\mathbf{Z}$, and **non-Archimedean** otherwise; the excuse for this terminology is that for an Archimedean valuation $\|m\|$ tends to infinity with m, and Archimedes wrote a book called *On Large Numbers*. It turns out that Archimedean and non-Archimedean valuations have significantly different properties.

Lemma 16 *If k is an algebraic number field, the Archimedean valuations on k are given by $\|\alpha\| = |\sigma\alpha|^c$ where $c > 0$ and σ is any embedding $k \to \mathbf{C}$.*

Proof The function on k defined in this way is certainly an Archimedean valuation, so let $\|.\|$ be any Archimedean valuation. It is clearly enough to prove the result when α is in $\mathfrak{o}$. Replacing $\|.\|$ by an equivalent valuation if necessary, we can assume that (21) holds. The first task is to prove

- $\|m\| = |m|^c$ for all m in $\mathbf{Z}$ and some fixed $c > 0$.

Clearly $\|\pm 1\|^2 = \|1\| = 1$, so we need only consider the case $m > 1$. Choose some $m_0 > 1$ and for any m, N write m^N in the scale of m_0:

$$m^N = \sum a_\nu m_0^\nu \quad \text{where } 0 \leqslant a_\nu < m_0. \tag{22}$$

Here the sum is taken over $0 \leqslant \nu \leqslant N \log m / \log m_0$. Let A be an upper bound for $\|a\|$ where $0 \leqslant a < m_0$. If we could choose m_0 with $\|m_0\| \leqslant 1$, applying (21) to (22) would give $\|m\|^N \leqslant A(1 + N \log m / \log m_0)$; taking N-th roots and letting $N \to \infty$ would then give $\|m\| \leqslant 1$ for all $m > 1$, contrary to hypothesis. Thus $\|m\| > 1$ for all $m > 1$. But now the same argument gives

$$\|m\|^N \leqslant A(1 + N \log m / \log m_0) \|m_0\|^{N \log m / \log m_0};$$

taking N-th roots and letting $N \to \infty$ we obtain $\|m\| \leqslant \|m_0\|^{\log m / \log m_0}$. If we had strict inequality here, we could interchange the roles of m_0 and m and obtain a contradiction. So $\|m\|^{1/\log m}$ is independent of m, and if we choose c so that it is equal to e^c our assertion follows.

Now let α be a non-zero element of $\mathfrak{o}$, and order the $\sigma_\nu : k \to \mathbf{C}$ so that $|\sigma_\nu \alpha| \geqslant |\sigma_{\nu+1} \alpha|$. (This ordering depends on α and should not be confused

with the ordering used in §3.) Let c have the value obtained in the previous
paragraph. Write

$$\prod_{\sigma}(X - \sigma\alpha^N) = X^n + a_1X^{n-1} + \cdots + a_n, \quad P_m = \prod_{\nu=1}^{\nu=m}(\sigma_\nu\alpha^N). \quad (23)$$

The a_m are symmetric functions of the $\sigma\alpha^N$ and the largest summand in
a_m is $\pm P_m$; so $|a_m| < C|P_m|$ where C depends only on n. Moreover, if
$|\sigma_{m+1}\alpha| < |\sigma_m\alpha|$ then once N is large enough this summand is much larger
than any other in a_m; so $|a_m| > \frac{1}{2}|P_m|$. Also $\|a_m\| = |a_m|^c$ because the a_m
are in $\mathbf{Z}$. If

$$|\sigma_\mu\alpha|^c > \|\alpha\| > |\sigma_{\mu+1}\alpha|^c$$

for some μ, then in the first equation (23) with α^N for X the term $a_\mu\alpha^{N(n-\mu)}$
on the right would be much larger than any of the others, which contradicts
the triangle inequality since the sum of all the terms vanishes. We get a sim-
ilar contradiction if $\|\alpha\| > |\sigma_1\alpha|^c$ or $\|\alpha\| < |\sigma_n\alpha|^c$; so $\|\alpha\| = |\sigma_\mu\alpha|^c$ for some
μ. Apparently μ here might depend on α; but using $\|\alpha\beta^N\| = \|\alpha\| \cdot \|\beta\|^N$
with N large we find first that we can require μ to be the same for β and
$\alpha\beta^N$ and then that it is the same for $\alpha\beta^N$ and α. $\square$

A refinement of this argument actually proves that any Archimedean
valuation on any field L is equivalent to one defined by $\|x\| = |\sigma x|$ for some
embedding $\sigma : L \to \mathbf{C}$. In our case there are $r_1 + r_2$ classes of Archimedean
valuations — one for each real embedding $k \to \mathbf{R}$ and one for each pair of
complex conjugate embeddings $k \to \mathbf{C}$. Note that $\mathbf{R}$ or $\mathbf{C}$ respectively can
be identified with the completion of k under the appropriate topology. The
Archimedean places are often called the **infinite places**, or by an abuse
of language motivated by Lemma 17 below, the infinite primes.

For non-Archimedean valuations we can radically improve (21), for if $\|.\|$
satisfies (21) and is non-Archimedean then

$$\|\alpha + \beta\|^N = \|\sum_{n=0}^{N} a_{n,N}\alpha^{N-n}\beta^n\| \leqslant \sum_{n=0}^{N}\|a_{n,N}\| \cdot \|\alpha^{N-n}\beta^n\|.$$

Here the $a_{n,N}$ are integers, so their absolute values are at most 1. The
right hand side is bounded by $(N+1)\max(\|\alpha\|, \|\beta\|)^N$. Taking N-th roots
and letting $N \to \infty$ gives

$$\|\alpha + \beta\| \leqslant \max(\|\alpha\|, \|\beta\|); \quad (24)$$

and if this holds for one valuation in an equivalence class, it holds for them
all. Applying (24) to $\alpha = (\alpha + \beta) + (-\beta)$ gives

$$\|\alpha\| \leqslant \max(\|\alpha + \beta\|, \|\beta\|);$$

and by symmetry there is equality in (24) whenever $\|\alpha\| \neq \|\beta\|$. A metric having this property is called an **ultrametric**.

Lemma 17 *Let $\|.\|$ be a non-Archimedean valuation. There are a non-zero prime ideal $\mathfrak{p}$ of $\mathfrak{o}$ and a constant $C > 1$ such that $\|\alpha\| = C^{-m}$ for all $\alpha \neq 0$ in k, where m is defined by $\mathfrak{p}^m \| \alpha$.*

Proof Any $\alpha \neq 0$ in $\mathfrak{o}$ satisfies an equation

$$\alpha^m + a_1 \alpha^{m-1} + \cdots + a_m = 0$$

where the a_μ are in $\mathbf{Z}$. If $\|\alpha\| > 1$ then the first term on the left would have strictly larger valuation than any of the others, which contradicts (24). So $\|\alpha\| \leqslant 1$ for all α in $\mathfrak{o}$. If $\|\alpha\| = 1$ for all $\alpha \neq 0$ in k then our valuation would be trivial; so there are integers $\alpha \neq 0$ with $\|\alpha\| < 1$. Using (24) again, the set of α with $\|\alpha\| < 1$ form an ideal $\mathfrak{p}$; and $\mathfrak{p}$ is prime because $\|\alpha_1 \alpha_2\| < 1$ implies $\|\alpha_1\| < 1$ or $\|\alpha_2\| < 1$.

Now choose π in $\mathfrak{p}$ but not in $\mathfrak{p}^2$ and let α be any non-zero element of k. In the notation of the lemma we can write $(\alpha/\pi^m) = \mathfrak{a}_1/\mathfrak{a}_2$ where $\mathfrak{a}_1, \mathfrak{a}_2$ are integral ideals prime to $\mathfrak{p}$. By Theorem 7 we can find β_2 in $\mathfrak{a}_2$ and prime to $\mathfrak{p}$. Write $\beta_1 = \beta_2 \alpha/\pi^m$, so that β_1 is in $\mathfrak{a}_1$. Neither β_1 nor β_2 is in $\mathfrak{p}$, so they both have valuation 1; thus $\|\alpha\| = \|\pi\|^m$. $\square$

Conversely, by the formula in the lemma any $\mathfrak{p}$ and C determine a non-Archimedean valuation; and changing C only changes the valuation within its place. This place can be identified with $\mathfrak{p}$, and will be called a **finite place**. Denote by $k_\mathfrak{p}$ the completion of k under the metric associated with $\mathfrak{p}$. An element α of $k_\mathfrak{p}$ is determined by a Cauchy sequence $\{\alpha_n\}$ where $\|\alpha_m - \alpha_n\| \to 0$ as $m, n \to \infty$; hence $\|\alpha_n\|$ tends to a limit, which we can define to be $\|\alpha\|$. It is easy to check that $k_\mathfrak{p}$ is a field containing k, and that $\|.\|$ determines a non-Archimedean valuation on $k_\mathfrak{p}$ which extends the given valuation on k. If we denote by $\mathfrak{o}_\mathfrak{p}$ the set of α in $k_\mathfrak{p}$ such that $\|\alpha\| \leqslant 1$ and by $\mathfrak{p}_\mathfrak{p}$ the set of α in $k_\mathfrak{p}$ such that $\|\alpha\| < 1$, then $\mathfrak{o}_\mathfrak{p}$ is an integral domain whose quotient field is $k_\mathfrak{p}$ and $\mathfrak{p}_\mathfrak{p}$ is its only non-zero prime ideal. Note that $\mathfrak{o}_\mathfrak{p}/\mathfrak{p}_\mathfrak{p}$ can be canonically identified with $\mathfrak{o}/\mathfrak{p}$. The elements of $\mathfrak{o}_\mathfrak{p}$ are in one-one correspondence with the nested sequences of residue classes

$$\alpha_1 \bmod \mathfrak{p} \supset \alpha_2 \bmod \mathfrak{p}^2 \supset \cdots \supset \alpha_m \bmod \mathfrak{p}^m \supset \cdots.$$

Now let the β_μ be a fixed set of representatives of the residue classes $\bmod \, \mathfrak{p}$ in $\mathfrak{o}$, and let π be in $\mathfrak{p}$ but not in $\mathfrak{p}^2$. (When we work with $k_\mathfrak{p}$ we shall consistently use π in this sense.) Thus the $\pi^{n-1}\beta_\mu$ are a set of representatives of the residue classes $\bmod \, \mathfrak{p}^n$ in $\mathfrak{p}^{n-1}$; hence the elements of $\mathfrak{o}_\mathfrak{p}$ are the $\sum_0^\infty \gamma_n \pi^n$ where each γ_n is a β_μ. Similarly the elements of $k_\mathfrak{p}$ are the $\sum_{-N}^\infty \gamma_n \pi^n$.

The field $k_\mathfrak{p}$ is an example of a **local field**, a subject for which there is a good introduction in [Ca] and a comprehensive account in [Se]. Analysis in $k_\mathfrak{p}$ is much easier than in $\mathbf{R}$; for example, it follows from (24) that a series in $k_\mathfrak{p}$ converges if and only if its terms tend to 0. The next three lemmas are further examples of this; they are stated only for $k_\mathfrak{p}$, but it will be clear from the proofs that they hold in much greater generality.

If we wish to refer to a place which is not necessarily finite, we usually denote it by v and denote the corresponding completion of k by k_v. If v is infinite, it is conventional to take $\mathfrak{o}_v$ to be the same as k_v.

The formulae for $\|.\|$ in Lemmas 16 and 17 each contain an arbitrary constant. But there is a canonical way of deriving a valuation on k_v in the class of $\|.\|$ from the induced topology on k. For any α in k_v, multiplication by α gives a map of k_v to itself which multiplies the natural measure of volume on k_v by a constant; we choose this constant to be $\|\alpha\|$. (The volume on k_v is only determined up to an arbitrary constant, but this does not affect the value of $\|\alpha\|$.) This process is also described, from a slightly different starting point, in §A1.3. For infinite places this gives $\|\alpha\| = |\sigma\alpha|$ if σ is real and $\|\alpha\| = |\sigma\alpha|^2$ otherwise; for the finite place associated with $\mathfrak{p}$ it gives $\|\alpha\| = (\mathrm{Norm}_k\mathfrak{p})^{-m}$ where $\mathfrak{p}^m$ is the exact power of $\mathfrak{p}$ which divides (α). This and $\mathrm{Norm}((\alpha)) = |\mathrm{norm}_{k/\mathbf{Q}}\alpha|$ immediately give the **Product Formula**

$$\prod_v \|\alpha\|_v = 1$$

where the product is taken over all normalized valuations.

Taking the logarithm of a non-Archimedean valuation and renormalizing, we obtain the **additive valuation** $\alpha \mapsto v_\mathfrak{p}(\alpha) = v$ where $\mathfrak{p}^v$ is the exact power of $\mathfrak{p}$ which divides α. (The double use of v is traditional, and should not cause confusion.) The corresponding rules for this are

$$v_\mathfrak{p}(\alpha\beta) = v_\mathfrak{p}(\alpha) + v_\mathfrak{p}(\beta), \quad v_\mathfrak{p}(\alpha + \beta) \geqslant \min(v_\mathfrak{p}(\alpha), v_\mathfrak{p}(\beta)),$$

and in the latter rule we have equality whenever the two arguments on the right are unequal.

Lemma 18 below says that under modest conditions factorizations of polynomials can be lifted from $\mathfrak{o}/\mathfrak{p}$ to $\mathfrak{o}_\mathfrak{p}$. This and Lemma 19 are special cases of **Hensel's Lemma**, for which the underlying idea is due to Newton. In its most general form (which belongs to algebraic geometry), this enables one to refine an approximate solution (in k or $k_\mathfrak{p}$) of a set of polynomial equations in k or $k_\mathfrak{p}$ to an exact solution in $k_\mathfrak{p}$. In Lemma 18 we shall again use a tilde to denote reduction $\mod \mathfrak{p}$.

Lemma 18 *Let $f(X)$ be a polynomial in $\mathfrak{o}_\mathfrak{p}[X]$ and $\tilde{f}(X) = \phi_1(X)\phi_2(X)$ where ϕ_1, ϕ_2 in $(\mathfrak{o}/\mathfrak{p})[X]$ are coprime. Then there exist polynomials f_1, f_2 in $\mathfrak{o}_\mathfrak{p}[X]$ with $f(X) = f_1(X)f_2(X)$, $\deg f_1 = \deg \phi_1$ and $\tilde{f}_\nu(X) = \phi_\nu(X)$.*

Proof We construct polynomials $f_1^{(n)}, f_2^{(n)}$ in $\mathfrak{o}_\mathfrak{p}[X]$ for $n = 1, 2, \ldots$ whose reductions $\mod \mathfrak{p}$ are ϕ_1, ϕ_2 and which have the properties $\deg f_1^{(n)} = \deg \phi_1$, $\deg f_2^{(n)} \leqslant \deg f - \deg \phi_1$,

$$\mathfrak{p}^n|(f - f_1^{(n)}f_2^{(n)}) \quad \text{and} \quad \mathfrak{p}^n|(f_\nu^{(n+1)} - f_\nu^{(n)}) \text{ for } \nu = 1, 2; \quad (25)$$

then the $f_\nu = \lim f_\nu^{(n)}$ will exist and have the required properties.

For the $f_\nu^{(1)}$ we lift the ϕ_ν to $\mathfrak{o}_\mathfrak{p}[X]$ in any way. To construct the $f_\nu^{(n+1)}$ from the $f_\nu^{(n)}$ we proceed as follows. By hypothesis

$$f = f_1^{(n)}f_2^{(n)} + \pi^n h^{(n)} \quad \text{for some } h^{(n)} \text{ in } \mathfrak{o}_\mathfrak{p}[X] \text{ with } \deg h^{(n)} \leqslant \deg f.$$

If we choose $f_\nu^{(n+1)} = f_\nu^{(n)} + \pi^n g_\nu^{(n)}$ with the $g_\nu^{(n)}$ in $\mathfrak{o}_\mathfrak{p}[X]$ then the second condition (25) will certainly be satisfied, and the first one will be equivalent to

$$h^{(n)} \equiv f_1^{(n)}g_2^{(n)} + f_2^{(n)}g_1^{(n)} \mod \mathfrak{p},$$

and therefore also to

$$\widetilde{h^{(n)}} = \phi_1 \widetilde{g_2^{(n)}} + \phi_2 \widetilde{g_1^{(n)}}.$$

Since ϕ_1, ϕ_2 are coprime and $(\mathfrak{o}/\mathfrak{p})[X]$ is a principal ideal domain, there are polynomials ψ_1, ψ_2 in $(\mathfrak{o}/\mathfrak{p})[X]$ such that $\phi_1\psi_2 + \phi_2\psi_1 = \widetilde{h^{(n)}}$ and $\deg \psi_1 < \deg \phi_1$; and we can take the $g_\nu^{(n)}$ to be any lifts of the ψ_ν. □

The next result is not quite so obvious as one would expect.

Corollary *Let K, k be algebraic number fields with $K \supset k$ and let $\mathfrak{P}, \mathfrak{p}$ be prime ideals in K, k respectively such that $\mathfrak{P}|\mathfrak{p}$. If α is in $\mathfrak{O}_\mathfrak{P}$ then α is integral over $\mathfrak{o}_\mathfrak{p}$; in particular $\mathrm{Tr}_{K_\mathfrak{P}/k_\mathfrak{p}}\alpha$ and $\mathrm{norm}_{K_\mathfrak{P}/k_\mathfrak{p}}\alpha$ are in $\mathfrak{o}_\mathfrak{p}$.*

Proof Suppose that $\mathfrak{P}^e\|\mathfrak{p}$ and choose Π in $\mathfrak{O}$ so that $\mathfrak{P}\|\Pi$. Let $B_1, \ldots, B_n$ be a base for $\mathfrak{O}/\mathfrak{P}$ as an $(\mathfrak{o}/\mathfrak{p})$-vector space. The representation on page 35 implies that the $\Pi^\mu B_\nu$ with $0 \leqslant \mu < e$ form a base for $\mathfrak{O}_\mathfrak{P}$ as an $\mathfrak{o}_\mathfrak{p}$-module. Hence $K_\mathfrak{P}$ is algebraic over $k_\mathfrak{p}$. In what follows, we shall use the absolute value associated with $\mathfrak{P}$, which clearly induces an absolute value on k associated with $\mathfrak{p}$. Let

$$f(X) = c_0 X^m + c_1 X^{m-1} + \cdots + c_m \quad (c_0 = 1)$$

be the minimal monic polynomial for α over $k_{\mathfrak{p}}$. We assume that the c_μ are not all in $\mathfrak{o}_{\mathfrak{p}}$ and obtain a contradiction. Let b in $\mathfrak{o}_{\mathfrak{p}}$ be such that the bc_μ are all in $\mathfrak{o}_{\mathfrak{p}}$ but not all divisible by $\mathfrak{p}$. If bc_m were the only one of the bc_μ not in $\mathfrak{p}$ then c_m would have strictly larger absolute value than any of the other terms in $f(\alpha) = 0$, contradicting the ultrametric law (24). In any other case, we can use the lemma to lift the factorization $\widetilde{bf}(X) \cdot 1$ to a non-trivial factorization of $bf(X)$ over $\mathfrak{o}_{\mathfrak{p}}$, and f would not be minimal. $\square$

It follows from Lemma 18 that if $\tilde{f}(X) = 0$ has a root of multiplicity 1 in $\mathfrak{o}/\mathfrak{p}$ then this root can be lifted to a root of $f(X) = 0$ in $\mathfrak{o}_{\mathfrak{p}}$. Here the multiplicity 1 condition is inconvenient, though we have to pay a price for dropping it.

Lemma 19 *Let $f(X)$ be a monic polynomial in $\mathfrak{o}_{\mathfrak{p}}[X]$ with formal derivative $f'(X)$, and let α in $\mathfrak{o}_{\mathfrak{p}}$ be such that $\|f(\alpha)\| < \|f'(\alpha)\|^2$. Then there is a unique root α^* of $f(X) = 0$ in $\mathfrak{o}_{\mathfrak{p}}$ such that*

$$\|\alpha^* - \alpha\| \leqslant \|f(\alpha)\|/\|f'(\alpha)\| < \|f'(\alpha)\|. \tag{26}$$

Proof We construct a sequence $\alpha_1 = \alpha, \alpha_2, \ldots$ of elements of $\mathfrak{o}_{\mathfrak{p}}$ such that

$$\left. \begin{array}{l} \|f(\alpha_{n+1})\| \leqslant \{\|f(\alpha)\|/\|f'(\alpha)\|^2\}\|f(\alpha_n)\| < \|f(\alpha_n)\|, \\[4pt] \|f'(\alpha_{n+1})\| = \|f'(\alpha)\|, \quad \|\alpha_{n+1} - \alpha_n\| \leqslant \|f(\alpha_n)\|/\|f'(\alpha)\|. \end{array} \right\} \tag{27}$$

These relations imply that $\|\alpha_{n+1} - \alpha_n\| \to 0$, so that the sequence tends to a limit α^* which is clearly a root of $f(X) = 0$.

If $\alpha_{n+1} = \alpha_n + \beta_n$ for some β_n in $\mathfrak{o}_{\mathfrak{p}}$ then

$$f(\alpha_{n+1}) = f(\alpha_n) + \beta_n f'(\alpha_n) + \text{ integral multiple of } \beta_n^2. \tag{28}$$

Take $\beta_n = -f(\alpha_n)/f'(\alpha_n)$, so that the first two terms on the right of (28) cancel. Now the penultimate claim in (27) follows from

$$f'(\alpha_{n+1}) = f'(\alpha_n) + \text{ integral multiple of } \beta_n$$

because the second term on the right has strictly smaller value than the first; and then the remaining claims follow at once.

It remains to prove uniqueness. But suppose α_1^*, α_2^* are distinct roots of $f(X) = 0$ satisfying (26); then $f(X) = (X - \alpha_1^*)(X - \alpha_2^*)g(X)$ with $g(X)$ in $\mathfrak{o}_{\mathfrak{p}}[X]$, whence $\|f'(\alpha_1^*)\| = \|\alpha_1^* - \alpha_2^*\| \cdot \|g(\alpha_1^*)\| \leqslant \|\alpha_1^* - \alpha_2^*\|$. $\square$

Some f are such that we can start this successive approximation process with a worse estimate α. An example which will be needed later is as follows.

Lemma 20 *Let p be the rational prime below $\mathfrak{p}$ and let α_0, ξ in $\mathfrak{o}_\mathfrak{p}^*$ be such that $\alpha_0^p \equiv \xi \bmod \mathfrak{p}^{m+r}$ where $\mathfrak{p}^m \| p$ and $r(p-1) > m$. Then there exists α in $\mathfrak{o}_\mathfrak{p}^*$ such that $\alpha^p = \xi$.*

Proof We again construct a sequence $\alpha_0, \alpha_1, \ldots$ such that

$$\xi \equiv \alpha_{n+1}^p \bmod \mathfrak{p}^{m+r+n+1}, \quad \alpha_{n+1} \equiv \alpha_n \bmod \mathfrak{p}^{n+r}; \tag{29}$$

then $\alpha = \lim \alpha_n$ will do what we want. Suppose that we have already chosen $\alpha_0, \ldots, \alpha_n$, and write $\alpha_{n+1} = \alpha_n + \pi^{n+r}\beta_n$ where β_n must be in $\mathfrak{o}_\mathfrak{p}$. The first congruence (29) will be satisfied if

$$\xi - \alpha_n^p \equiv p\pi^{n+r}\alpha_n^{p-1}\beta_n \bmod \mathfrak{p}^{m+r+n+1}$$

and this is always possible. $\square$

Lemma 21 (i) *Let $L = k_\mathfrak{p}(\alpha)$ be a finite algebraic extension of $k_\mathfrak{p}$ and let $f(X)$ be the minimal monic polynomial for α over $k_\mathfrak{p}$. If $g(X)$ in $k_\mathfrak{p}[X]$ is monic and close enough to $f(X)$, then there exists β in L such that $g(\beta) = 0$ and $L = k_\mathfrak{p}(\beta)$, and $g(X)$ is irreducible over $k_\mathfrak{p}$.*

(ii) *Every finite algebraic extension of $k_\mathfrak{p}$ lies in some $K_\mathfrak{P}$ where K is a finite algebraic extension of k and $\mathfrak{P}$ is a prime of K above $\mathfrak{p}$.*

Proof To prove (i), we would like to apply Lemma 19 to α as an approximate solution of $g(X) = 0$; but this is illegitimate because we do not yet have a valuation on $k_\mathfrak{p}(\alpha)$. Instead we replace $f(X)$ by a polynomial $g_0(X)$ close to $f(X)$ and in $\mathfrak{o}[X]$; and if β is a root of g_0 we choose a neighbourhood $\mathcal{N}$ of f in the space of all monic polynomials in $k_\mathfrak{p}[X]$ with the same degree as f so that the conditions of Lemma 19 hold for β and any g in $\mathcal{N}$. For $k_\mathfrak{p}(\beta)$ to be meaningful we must embed $k_\mathfrak{p}$ and β in a common field; it will be convenient to take this field to be $K_\mathfrak{P}$, where K is the splitting field of $g_0(X)$ over k and $\mathfrak{P}$ is any prime of K lying above $\mathfrak{p}$. Clearly the valuation on K associated with $\mathfrak{P}$ restricts to a valuation on k associated with $\mathfrak{p}$, though the latter may not be canonically normalized; thus we can use $\|.\|$ to denote both of them.

After multiplying α by a suitable element of $\mathfrak{o}$, we can assume that $f(X)$ is in $\mathfrak{o}_\mathfrak{p}[X]$. Define $D(g_1, g_2) = \prod g_1'(\xi_i)$ for any monic polynomials g_1, g_2 in $k_\mathfrak{p}[X]$ and of the same degree as f, where the product is taken over all roots ξ_i of g_2; thus $D(g_1, g_2)$ is equal to a polynomial in the coefficients of g_1 and g_2, and $c = \|D(f, f)\| \neq 0$ because f has no repeated roots. We require $\mathcal{N}$ to be so small that $\|D(g_1, g_2)\| = c$ for all g_1, g_2 in $\mathcal{N}$ and that every g in $\mathcal{N}$ is in $\mathfrak{o}_\mathfrak{p}[X]$. We impose the further condition on $\mathcal{N}$ that for g in $\mathcal{N}$ every coefficient of $f - g$ has absolute value strictly less than c^2. If we take g_0 to be in $\mathcal{N} \cap \mathfrak{o}[X]$ then every coefficient of $g - g_0$ admits the same bound and hence

$$\|g(\beta)\| = \|g(\beta) - g_0(\beta)\| < c^2.$$

If β_i is any root of g_0, then $g'(\beta_i)$ is in $\mathfrak{O}_\mathfrak{P}$; so $\|g'(\beta_i)\| \leqslant 1$ and therefore $\|g'(\beta)\| \geqslant c$. Hence the hypotheses of Lemma 19 are satisfied, so to each root β_i of g_0 there corresponds a root γ_i of g in $\mathfrak{O}_\mathfrak{P}$ with $\|\gamma_i - \beta_i\| < c$. But if β_i, β_j are distinct roots of g_0 then $\|\beta_i - \beta_j\| \geqslant \|g_0'(\beta_i)\| \geqslant c$; so if also $\|\gamma_j - \beta_j\| < c$ then γ_i and γ_j are distinct. Since g has only as many roots as g_0, each root of g must occur as a γ_i and is therefore in $\mathfrak{O}_\mathfrak{P}$. Moreover $k_\mathfrak{p}(\gamma_i) \subset k_\mathfrak{p}(\beta_i) \subset K_\mathfrak{P}$ for some root β_i of g_0.

Taking g to be f, this proves (ii). Knowing this, we can take g_0 in the argument above to be any polynomial in $\mathcal{N}$. But $k_\mathfrak{p}(\beta) \supset k_\mathfrak{p}(\alpha)$, so that in

$$\deg g \geqslant [k_\mathfrak{p}(\beta) : k_\mathfrak{p}] \geq [k_\mathfrak{p}(\alpha) : k_\mathfrak{p}] = \deg f$$

the two outer terms are equal; it follows that $k_\mathfrak{p}(\beta) = k_\mathfrak{p}(\alpha)$ and $g(X)$ is irreducible over $k_\mathfrak{p}$. $\square$

In the language which we have introduced in this section, we can restate the Chinese Remainder Theorem (Theorem 7) as follows. Let $\mathfrak{p}_1, \ldots, \mathfrak{p}_m$ be distinct prime ideals in $\mathfrak{o}$ and $\alpha_1, \ldots, \alpha_m$ any elements of $\mathfrak{o}$; for any $\epsilon > 0$ we can find α in $\mathfrak{o}$ such that $\|\alpha - \alpha_\mu\|_{\mathfrak{p}_\mu} < \epsilon$ for each μ. A variant of this which does not discriminate against the infinite places is the **Weak Approximation Theorem**:

Theorem 17 *Let $v_1, \ldots, v_m$ be distinct places of k and $\alpha_1, \ldots, \alpha_m$ any elements of k. To any $\epsilon > 0$ we can find α in k such that $\|\alpha - \alpha_\mu\|_{v_\mu} < \epsilon$ for each μ.*

Proof We saw on page 3 that k is dense in $\prod k_v$, where the product is over all infinite places; so we can find β in k satisfying $\|\beta - \alpha_\mu\|_{v_\mu} < \frac{1}{2}\epsilon$ for all the infinite places among the v_μ. Choose $M > 0$ in $\mathbf{Z}$ such that $M(\alpha_\mu - \beta)$ is in $\mathfrak{o}$ for each finite place v_μ. By Theorem 7 we can find γ in $\mathfrak{o}$ such that

$$\|\gamma - M(\alpha_\mu - \beta)\|_{v_\mu} < \tfrac{1}{2}\epsilon\|M\|_{v_\mu}$$

for each finite place v_μ; then $\alpha = \gamma/MN + \beta$ will satisfy all our conditions provided that $N > 0$ in $\mathbf{Z}$ satisfies

$$\|N\|_{v_\mu} > 2\epsilon^{-1}\|\gamma/M\|_{v_\mu} \quad \text{for each infinite place } v_\mu, \qquad (30)$$

$$\|N - 1\|_{v_\mu} < \tfrac{1}{2}\epsilon\|M/\gamma\|_{v_\mu} \qquad \text{for each finite place } v_\mu. \qquad (31)$$

For this, it is enough to choose N so that $N-1$ is divisible by a large power of Norm $\mathfrak{p}_\mu$ for every $\mathfrak{p}_\mu$ associated with a finite place v_μ. $\qquad\square$

Exercise In the notation of the theorem, prove by induction on m that there exists β in k such that $\|\beta\|_{v_1} > 1$ and $\|\beta\|_{v_\mu} < 1$ for $\mu = 2, \ldots, m$. [Use the proof of Lemma 15 to find γ with $\|\gamma\|_{v_1} > 1$ and $\|\gamma\|_{v_m} < 1$. By the induction hypothesis there exists δ with $\|\delta\|_{v_1} > 1$ and $\|\delta\|_{v_\mu} < 1$ for $\mu = 2, \ldots, m - 1$. Now take $\beta = \gamma\delta^r$ if $\|\delta\|_{v_m} \leqslant 1$ or $\beta = \gamma\delta^r/(1 + \delta^r)$ if $\|\delta\|_{v_m} > 1$, where r is large.] By considering $\beta^r/(1 + \beta^r)$ for r large, deduce the theorem in the special case $\alpha_1 = 1$, $\alpha_2 = \cdots = \alpha_m = 0$ and hence derive it in general. $\qquad\square$

The significance of the phrase 'weak approximation' will be explained in §9, where Theorem 17 will be translated into the language of adèles. Roughly speaking, a weak approximation theorem asserts the existence of one or more elements of k satisfying certain conditions, whereas a strong approximation theorem asserts the existence of one or more elements of $\mathfrak{o}$, or $\mathfrak{o}_S$ for a preassigned finite set S of prime ideals, satisfying certain conditions. (Thus Theorem 7 is a strong approximation theorem.) Strong approximation theorems are in general much harder to prove than weak approximation ones, and are usually uglier to state. For example, the strong analogue of Theorem 17 is trivially false for $\mathfrak{o}$. If we add the further condition that there is at least one infinite place which is not among the v_μ, we obtain a plausible but unproved conjecture.

7 Field extensions and ramification

Let $K = k(\alpha)$ be an algebraic number field with $[K : k] = n$, let $\phi(X)$ be the minimal monic polynomial for α over k, let $\mathfrak{p}$ be a prime ideal in $\mathfrak{o}_k$ and let p be the rational prime divisible by $\mathfrak{p}$. There is a close relationship between the factorization of ϕ in $k_{\mathfrak{p}}$ and the factorization of $\mathrm{conorm}_{K/k}\mathfrak{p}$ in $\mathfrak{O}_K$. The latter must have the form

$$\mathrm{conorm}_{K/k}\mathfrak{p} = \prod \mathfrak{P}_\mu^{e_\mu} \qquad (32)$$

for some primes $\mathfrak{P}_1, \dots, \mathfrak{P}_g$ in $\mathfrak{O}_K$. We say that $\mathfrak{p}$ **ramifies** in K/k if some $e_\mu > 1$; the ramification is called **wild** if some e_μ is divisible by p, and **tame** otherwise. If we write $\mathrm{norm}_{K/k}\mathfrak{P}_\mu = \mathfrak{p}^{f_\mu}$, the Corollary to Lemma 13 gives

$$n = [K:k] = \sum e_\mu f_\mu. \tag{33}$$

We temporarily drop the subscript μ. Let $\beta_1, \dots, \beta_f$ in $\mathfrak{O}_K$ be such that their images are a base for $\mathfrak{O}/\mathfrak{P}$ as an $(\mathfrak{o}/\mathfrak{p})$-vector space, and let Π in $\mathfrak{O}_K$ be such that $\mathfrak{P}\|\Pi$. Then the $\beta_i\Pi^j$ with $0 \leqslant j < e$ form a base for $\mathfrak{O}_K/\mathfrak{P}$ as an $(\mathfrak{o}/\mathfrak{p})$-vector space, so that $[K_\mathfrak{P} : k_\mathfrak{p}] = ef$. We can however strengthen this result considerably:

Theorem 18 *In the notation above, there is a natural isomorphism*

$$K \otimes_k k_\mathfrak{p} \approx K_{\mathfrak{P}_1} \oplus \cdots \oplus K_{\mathfrak{P}_g} \tag{34}$$

both algebraically and topologically.

Proof By (33), both sides of (34) are $k_\mathfrak{p}$-vector spaces of the same finite dimension. There are natural maps $K \otimes_k k_\mathfrak{p} \to K_{\mathfrak{P}_\mu}$ and hence there is a continuous vector space homomorphism ϕ from the left hand side of (34) to the right hand side. But $\phi(K)$ is dense in the right hand side, by Theorem 17, so ϕ is onto. $\qquad\square$

Theorem 19 *In the notation above, let $\phi(X) = \prod \phi_\mu(X)$ where the $\phi_\mu(X)$ are irreducible in $k_\mathfrak{p}[X]$. Then after renumbering, $\deg \phi_\mu = e_\mu f_\mu$ and $\phi_\mu(X)$ is the minimal monic polynomial for α over $K_{\mathfrak{P}_\mu}$.*

Proof For each fixed μ, it follows from Theorem 18 that α generates $K_{\mathfrak{P}_\mu}$ over $k_\mathfrak{p}$. Let $\phi_\mu(X)$ be the minimal monic polynomial for α over $k_\mathfrak{p}$ when both are considered as lying in $K_{\mathfrak{P}_\mu}$. (This will depend on μ; the point is that we cannot define binary operations between α and elements of $k_\mathfrak{p}$ until we have embedded them both in a common field.) Thus $\phi_\mu(X)$ has degree $[K_\mathfrak{P} : k_\mathfrak{p}] = e_\mu f_\mu$ and divides $\phi(X)$; comparing degrees and using (33), it only remains to show that the $\phi_\mu(X)$ are distinct. But the topology on $K_{\mathfrak{P}_\mu}$ is determined by a knowledge of $\phi_\mu(X)$, and by Theorem 17 the topologies induced on K by the various $\mathfrak{P}_\mu$ are all different. $\qquad\square$

We shall shortly see that when $k = \mathbf{Q}$ the p with some $e > 1$ are just those which divide d_K; hence for any k the primes $\mathfrak{p}$ in $\mathfrak{o}$ which ramify in K/k also divide d_K. Assume for convenience that we have chosen α to be an integer. In the notation above $\phi(X)$ is in $\mathfrak{o}[X]$ and if one excludes finitely many p then $\tilde{\phi}(X) = 0$ has no repeated roots. If so, by Lemma

18 the factorization of $\phi(X)$ over $k_\mathfrak{p}$ corresponds to that of $\tilde{\phi}(X)$ over $\mathbf{F}_q$ where $q = \operatorname{Norm}\mathfrak{p}$; and if $\phi_1(X)$ corresponds to $\mathfrak{P}_1$ then we can take $\mathfrak{P}_1 = (\mathfrak{p}, \psi_1(\alpha))$ where $\psi_1(X)$ is any lift of $\tilde{\phi}_1(X)$ to $\mathfrak{o}[X]$. We can actually do better than this.

Lemma 22 *With the notation above, let α in $\mathfrak{O}$ be such that the index of $\mathfrak{o}[\alpha]$ in $\mathfrak{O}$ is finite and prime to $\mathfrak{p}$. Then $\tilde{\phi}(X) = \prod(\psi_\mu(X))^{e_\mu}$ where the $\psi_\mu(X)$ are in $\mathbf{F}_q[X]$, coprime and irreducible, and $\deg \psi_\mu = f_\mu$. If $g_\mu(X)$ is any lift of $\psi_\mu(X)$ to $\mathfrak{o}[X]$ then $\mathfrak{P}_\mu = (\mathfrak{p}, g_\mu(\alpha))$.*

Proof Let $\psi_1(X)$ be a monic irreducible factor of $\tilde{\phi}(X)$ over $\mathbf{F}_q$ and let β in $\overline{\mathbf{F}_q}$ be a root of $\psi_1(X)$. Every element of $\mathfrak{O}$ can be written as $\sum_0^{n-1} c_\nu \alpha^\nu$ where the c_ν are in k with denominators prime to $\mathfrak{p}$ and $n = [K : k]$; hence there is an epimorphism $\mathfrak{O} \to \mathbf{F}_q[\beta]$ given by $\alpha \mapsto \beta$ which extends reduction mod $\mathfrak{p}$. Let $\mathfrak{P}$ be its kernel; since the image is a field, $\mathfrak{P}$ is a prime ideal which divides $\mathfrak{p}$. If e, f denote the standard values corresponding to $\mathfrak{P}$ then $\deg \psi_1 = f$ because $\mathbf{F}_q[\beta] \approx \mathfrak{O}/\mathfrak{P}$ has order $(\operatorname{Norm}\mathfrak{p})^f$. Moreover $\mathfrak{P} \supset (\mathfrak{p}, g_1(\alpha))$. Conversely any element γ of $\mathfrak{P}$ has the form $h(\alpha)$ where $h(X)$ is in $k[X]$ and the coefficients of h have denominators prime to $\mathfrak{p}$; and $\tilde{h}(\beta) = 0$ so that ψ_1 divides $\tilde{h}$. In other words $h(X) - g_1(X)h_1(X)$ has all its coefficients in $\mathfrak{p}$ for some $h_1(X)$ in $\mathfrak{o}[X]$; setting $X = \alpha$ we deduce that γ is in $(\mathfrak{p}, g_1(\alpha))$. In particular $\mathfrak{P} \| g_1(\alpha)$ if $e > 1$.

We have $\tilde{\phi}(X) = \prod(\psi_\mu(X))^{c_\mu}$ for some c_μ. Thus $\phi(X) - \prod(g_\mu(X))^{c_\mu}$ has all its coefficients in $\mathfrak{p}$, so $\prod(g_\mu(\alpha))^{c_\mu}$ lies in $\mathfrak{p}$. On the other hand $g_\mu(\alpha)$ maps to $\psi_\mu(\beta) \neq 0$ if $\mu \neq 1$, so $g_\mu(\alpha)$ is not in $\mathfrak{P}$ and the distinct g_μ correspond to distinct prime factors of $\mathfrak{p}$ in K. Hence $(g_1(\alpha))^{c_1}$ is divisible by $\mathfrak{P}^e$; so $c_1 \geqslant e$ if $e > 1$, by the last result in the previous paragraph. This also holds if $e = 1$. Comparing $[K : k] = \sum c_\mu f_\mu$ with (33) shows that $c_1 = e$ and that each $\mathfrak{P}_\mu$ is generated in this way.   □

The most favourable situation is when $\mathfrak{O} = \mathfrak{o}[\alpha]$, which seems to happen quite often. (There is a problem with assertions like this, because it is not clear what one should mean by an arbitrary algebraic number field.) But there are pairs $K, \mathfrak{p}$ for which there does not exist an α satisfying the conditions of Lemma 22, though this can only happen if $\operatorname{Norm}\mathfrak{p} \leqslant [K : k]$. (See Exercise 2.8.) In such cases there may still be labour-saving tricks available, but sometimes the least onerous way to find how $\mathfrak{p}$ factorizes is to study $\mathfrak{O}/\mathfrak{p}$. However, the following lemma, which is needed for a quite different reason, does enable one to modify Lemma 22 so as to compute the $\mathfrak{P}_\mu$ and their associated e_μ, f_μ one by one.

Lemma 23 *Let* $\mathfrak{p}, \mathfrak{P}$ *be prime ideals in algebraic number fields* k, K *respectively with* $K \supset k$ *and* $\mathfrak{P}|\mathfrak{p}$. *There exist an algebraic integer* α *in* K *and embeddings of* $\mathfrak{o}_\mathfrak{p}$ *and* α *into* $K_\mathfrak{P}$ *such that* $K_\mathfrak{P} = k_\mathfrak{p}(\alpha)$ *and* $\mathfrak{o}_\mathfrak{p}[\alpha] = \mathfrak{O}_\mathfrak{P}$.

Proof Let ξ generate $\mathfrak{O}/\mathfrak{P}$ over $\mathfrak{o}/\mathfrak{p}$ and lift ξ back to an integer α in $\mathfrak{O}$; then $\mathfrak{o}_\mathfrak{p}[\alpha]$ contains representatives of all the classes in $\mathfrak{O}/\mathfrak{P}$. If it also contains an element divisible by $\mathfrak{P}$ only to the first power, then it contains representatives of every class in $\mathfrak{P}^m/\mathfrak{P}^{m+1}$ for each $m > 0$, and the lemma follows at once. But let $\phi(X)$ in $\mathfrak{o}[X]$ be such that $\tilde{\phi}(X)$ is the minimal monic polynomial for ξ over $\mathfrak{o}/\mathfrak{p}$, and let $\phi'(X)$ be the formal derivative of $\phi(X)$. Here $\widetilde{\phi'}(X)$ does not vanish identically because $\mathfrak{O}/\mathfrak{P}$ is separable over $\mathfrak{o}/\mathfrak{p}$; hence $\widetilde{\phi'}(\xi) \neq 0$ because $\tilde{\phi'}$ has smaller degree than $\tilde{\phi}$. If Π in $\mathfrak{O}$ is such that $\mathfrak{P}\|\Pi$ then $\mathfrak{P}|\phi(\alpha)$ and

$$\phi(\alpha + \Pi) \equiv \phi(\alpha) + \Pi\phi'(\alpha) \bmod \mathfrak{P}^2. \tag{35}$$

By (35) at least one of α and $\alpha + \Pi$ satisfies the condition in the second sentence of the proof, and they are both lifts of ξ. $\qquad\square$

Exercise Obtain the analogues of Theorems 18 and 19 for infinite places.

8 The Different

Many number-theoretic objects give rise to an ideal which identifies the bad primes for that object and measures how bad they are. Such an ideal is important primarily because it has this property, but the definition does not always make this evident; on occasion this fact can be a non-trivial theorem. Typically such an ideal is called the **conductor**, but for an extension K/k of algebraic number fields it is called the (relative) **different**.

By Lemma 3, every k-linear map $K \to k$ has the form $\hat{\alpha} \mapsto \text{Tr}_{K/k}(\alpha\beta)$ for some β in K. It is natural to ask which are the β for which the image of $\mathfrak{O}_K$ under this map lies in $\mathfrak{o}_k$. Let $\mathcal{S}$ be the set of such β; clearly $\mathcal{S}$ is an $\mathfrak{O}_K$-module which contains $\mathfrak{O}_K$. On the other hand, $\text{Tr}_{K/\mathbf{Q}}(\alpha\beta)$ must lie in $\mathbf{Z}$. So if $\alpha_1, \ldots, \alpha_N$ are a base for $\mathfrak{O}_K$ as a $\mathbf{Z}$-module then $\mathcal{S}$ must lie in the $\mathbf{Z}$-module spanned by $\beta_1, \ldots, \beta_N$ where β_ν satisfies

$$\text{Tr}(\alpha_\mu\beta_\nu) = \begin{cases} 1 & \text{if } \mu = \nu, \\ 0 & \text{otherwise.} \end{cases} \tag{36}$$

If we write $\beta_\nu = \sum \alpha_\rho c_{\rho\nu}$ with $c_{\rho\nu}$ in $\mathbf{Q}$ then we have the matrix relation

$$(\text{Tr}(\alpha_\mu\alpha_\rho))((c_{\rho\nu})) = I.$$

It follows that each $d_K c_{\rho\nu}$ is in $\mathbf{Z}$ where d_K is the discriminant of K, so

S is contained in the fractional ideal $d_K^{-1}\mathfrak{O}_K$; since S is an $\mathfrak{O}_K$-module, it is itself a fractional ideal. Since also $S \supset \mathfrak{O}_K$, it is natural to write S as $\mathfrak{d}^{-1} = \mathfrak{d}_{K/k}^{-1}$ where $\mathfrak{d}_{K/k}$ is an integral ideal called the **different** of K/k. If $k = \mathbf{Q}$, the β_ν defined above span $\mathfrak{d}_{K/\mathbf{Q}}^{-1}$ and it follows that

$$\text{Norm}\,\mathfrak{d}_{K/\mathbf{Q}} = |\det(\text{Tr}_{K/\mathbf{Q}}(\alpha_\mu \alpha_\rho))| = |d_K|. \tag{37}$$

Now let $K \supset L \supset k$ and let β run through the elements of $\mathfrak{O}_K$, λ through the elements of $\mathfrak{O}_L$ and γ through the elements of $\mathfrak{d}_{L/k}$. If α is in K then

$$\alpha \text{ in } \mathfrak{d}_{K/k}^{-1} \Longleftrightarrow \text{Tr}_{K/k}(\alpha\beta) \text{ is in } \mathfrak{o}_k \text{ for all } \beta$$
$$\Longleftrightarrow \text{Tr}_{K/k}(\alpha\beta\lambda) = \text{Tr}_{L/k}(\lambda\text{Tr}_{K/L}(\alpha\beta)) \text{ is in } \mathfrak{o}_k \text{ for all } \beta, \lambda$$
$$\Longleftrightarrow \text{Tr}_{K/L}(\alpha\beta) \text{ is in } \mathfrak{d}_{L/k}^{-1} \text{ for all } \beta$$
$$\Longleftrightarrow \gamma\text{Tr}_{K/L}(\alpha\beta) = \text{Tr}_{K/L}(\alpha\gamma\beta) \text{ is in } \mathfrak{O}_L \text{ for all } \beta, \gamma$$
$$\Longleftrightarrow \alpha\gamma \text{ is in } \mathfrak{d}_{K/L}^{-1} \text{ for all } \gamma \Longleftrightarrow \alpha \text{ in } \mathfrak{d}_{K/L}^{-1}\mathfrak{d}_{L/k}^{-1};$$

and therefore

$$\mathfrak{d}_{K/k} = \mathfrak{d}_{K/L}\mathfrak{d}_{L/k}. \tag{38}$$

Let $\mathfrak{P}$ be a prime ideal of K and let $\mathfrak{p}$ be the prime ideal of k which it divides. By analogy with what we have already done globally, we can consider the set of β in $K_\mathfrak{P}$ such that $\text{Tr}_{K_\mathfrak{P}/k_\mathfrak{p}}(\alpha\beta)$ is in $\mathfrak{o}_\mathfrak{p}$ whenever α is in $\mathfrak{O}_\mathfrak{P}$. As in the global case, this set is a fractional ideal in $K_\mathfrak{P}$; we again write it as $\mathfrak{d}^{-1}$ and we now call $\mathfrak{d}$ the **local different** for the extension $K_\mathfrak{P}/k_\mathfrak{p}$. We identify $\mathfrak{d}$ with a power of $\mathfrak{P}$, though strictly speaking it is a power of $\mathfrak{P}_\mathfrak{P}$.

Lemma 24 *The global different $\mathfrak{d}_{K/k}$ is the product of the local differents.*

Proof Suppose that η generates K over k; then by Theorem 19

$$\text{Tr}_{K/k}\eta = \sum_\mathfrak{P} \text{Tr}_{K_\mathfrak{P}/k_\mathfrak{p}}\eta \quad \text{for } \eta \text{ in } K. \tag{39}$$

But any η in K can be written as $\eta' - \eta''$ where each of η' and η'' generates K over k, and (39) holds for η because it holds for η' and η''. Now let $\mathfrak{p}$ be any prime ideal of k. It follows from (39) that if β in K lies in $\mathfrak{d}_{K_\mathfrak{P}/k_\mathfrak{p}}^{-1}$ for each $\mathfrak{P}$ above $\mathfrak{p}$ and if α is in $\mathfrak{O}_K$ then $\text{Tr}_{K/k}(\alpha\beta)$ is integral at $\mathfrak{p}$. By considering all such β which are integral at every prime of K which does not divide $\mathfrak{p}$, we conclude that $\prod \mathfrak{d}_{K_\mathfrak{P}/k_\mathfrak{p}}$ divides $\mathfrak{d}_{K/k}$ where the product is taken over all $\mathfrak{P}$ dividing $\mathfrak{p}$.

Conversely, suppose that $\mathfrak{P}^r \| \mathfrak{d}_{K/k}$ and let β in K be in $\mathfrak{P}^{-r}$ but not in

$\mathfrak{P}^{1-r}$. For any α in $\mathfrak{O}_K$ it follows from (39) that $\text{Tr}_{K_{\mathfrak{P}}/k_{\mathfrak{p}}}(\alpha\beta)$ is in $\mathfrak{o}_{\mathfrak{p}}$, for all the other terms in (39) are certainly in $\mathfrak{o}_{\mathfrak{p}}$. Thus β is in $\mathfrak{d}_{K_{\mathfrak{P}}/k_{\mathfrak{p}}}$, whence $\mathfrak{P}^r | \mathfrak{d}_{K_{\mathfrak{P}}/k_{\mathfrak{p}}}$. $\qquad\square$

The proof of Theorem 20 below is complicated. The motivation behind it is that if we write $[K : k] = n$, we can find $\mathfrak{d}_{k/k}^{-1}$ by taking a base $\alpha_1, \ldots, \alpha_n$ for $\mathfrak{O}_K$ as an $\mathfrak{o}_k$-module and solving the equations (36) for the β_ν, where now the trace is for K/k. If we could choose the α_ν to be $1, \alpha, \ldots, \alpha^{n-1}$ for some integer α with $K = k(\alpha)$, there would be a sensible formula for the solution of (36). In the first part of the proof we derive such a formula; instead of using brute force, we apply a trick which is worth knowing because it turns up in a variety of contexts. For the global extension K/k we cannot expect to have such a base. But we showed in Lemma 23 that for each prime $\mathfrak{P}$ in K we can find an α which has the corresponding local property; and using Lemma 24 this is good enough.

Theorem 20 *Let α run through those integers of K for which $K = k(\alpha)$ and let $\phi(X)$ be the minimal monic polynomial of α over k; then $\mathfrak{d}_{K/k}$ is the highest common factor of the $\phi'(\alpha)$.*

Proof Write

$$\frac{\phi(X)}{X - \alpha} = \beta_0 X^{n-1} + \beta_1 X^{n-2} + \cdots + \beta_{n-1},$$

so that the β_ν are in $\mathfrak{o}_k[\alpha]$ and $\beta_0 = 1$. In the algebraic closure $\bar{k}$

$$\sum \frac{\phi(X)}{X - \alpha_\nu} \frac{\alpha_\nu^\mu}{\phi'(\alpha_\nu)} = X^\mu \quad \text{for} \quad 0 \leqslant \mu \leqslant n - 1$$

where the sum is over the roots α_ν of $\phi(X) = 0$; for the difference of the two sides is a polynomial of degree at most $n - 1$ which vanishes at the n values $X = \alpha_\nu$. Equating coefficients of X^ρ gives

$$\text{Tr}_{K/k} \left(\alpha^\mu \frac{\beta_{n-1-\rho}}{\phi'(\alpha)} \right) = \begin{cases} 1 & \text{if } \mu = \rho, \\ 0 & \text{otherwise}; \end{cases}$$

hence the $\beta_{n-1-\rho}/\phi'(\alpha)$ are linearly independent over k and $\text{Tr}(\alpha^\mu\beta)$ is in $\mathfrak{o}_k$ for $\mu = 0, 1, \ldots, n - 1$ if and only if β is in the $\mathfrak{o}_k$-module S spanned by the $\beta_{n-1-\rho}/\phi'(\alpha)$. It follows that $S \supset \mathfrak{d}_{K/k}^{-1}$; however, $S \subset (1/\phi'(\alpha))$, so $\mathfrak{d}_{K/k} \supset (\phi'(\alpha))$ for each α. If $1, \alpha, \ldots, \alpha^{n-1}$ are a base for $\mathfrak{O}_K$ as an $\mathfrak{o}_k$-module, this argument gives $\mathfrak{d}_{K/k} = (\phi'(\alpha))$ because $\beta_0 = 1$.

By Lemma 23 with $\mathfrak{P} = \mathfrak{P}_1$, we can choose α so that $\mathfrak{o}_{\mathfrak{p}}[\alpha] = \mathfrak{O}_{\mathfrak{P}_1}$. The construction ensures that α is not in $\mathfrak{P}_1$, and it allows us to require that α

is in $\mathfrak{P}_\mu$ for all $\mu > 1$. Hence in the notation of Theorem 19, for each $\mu > 1$ the constant term of $\phi_\mu(X)$ is divisible by $\mathfrak{P}_\mu$ and hence by $\mathfrak{p}$. Since $\phi_\mu(X)$ is irreducible in $k_\mathfrak{p}$, it follows from Lemma 18 that all the coefficients in $\phi_\mu(X)$ except the leading one are divisible by $\mathfrak{p}$; and hence $\phi_\mu(\alpha)$ is not divisible by $\mathfrak{P}_1$. This implies

$$\phi'(\alpha) \text{ and } \phi_1'(\alpha) \text{ are divisible by the same power of } \mathfrak{P}_1. \qquad (40)$$

But $\phi_1(X)$ is the minimal monic polynomial for α over $k_\mathfrak{p}$, where α is regarded as an element of $K_{\mathfrak{P}_1}$; so an argument like that in the previous paragraph shows that the local different for $K_{\mathfrak{P}_1}/k_\mathfrak{p}$ is equal to $(\phi_1'(\alpha))$. The theorem now follows from (40) and Lemma 24. $\square$

Theorem 21 *Let $\mathfrak{p}, \mathfrak{P}$ be prime ideals in algebraic number fields k, K respectively with $K \supset k$ and let $e > 0$ be given by $\mathfrak{P}^e \| \mathfrak{p}$. Then $\mathfrak{P}^{e-1} | \mathfrak{d}_{K/k}$; and $\mathfrak{P}^e | \mathfrak{d}_{K/k}$ if and only if $\mathfrak{p} | e$.*

Proof Suppose temporarily that α is in $\mathfrak{P}_\mu$ in the notation of Theorem 19. Then the constant term in $\phi_\mu(X)$ must be divisible by $\mathfrak{P}_\mu$ and hence by $\mathfrak{p}$. As in the proof of Theorem 20 every coefficient of $\phi_\mu(X)$ except the leading one must be divisible by $\mathfrak{p}$; for otherwise $\phi_\mu(X)$ would be reducible, by Lemma 18. In particular

$$\alpha \text{ in } \mathfrak{P} \quad \Longrightarrow \quad \mathfrak{p} \text{ divides } \mathrm{Tr}_{K_\mathfrak{P}/k_\mathfrak{p}}(\alpha). \qquad (41)$$

Now let β be any element of $\mathfrak{P}^{1-e}$; if $\mathfrak{p} \| \pi$ then $\pi\beta$ is in $\mathfrak{P}_\mu$ for every μ, so that (39) and (41) imply $\mathfrak{p} | \mathrm{Tr}_{K/k}(\pi\beta)$ whence $\mathrm{Tr}_{K/k}(\beta)$ is in $\mathfrak{o}_\mathfrak{p}$. Thus $\mathfrak{P}^{e-1}$ divides $\mathfrak{d}_{K/k}$.

Now revert to the assumption that α is in $\mathfrak{O}$, and denote reduction mod $\mathfrak{P}$ by a tilde. Then $\tilde{\alpha}$ is a root of $\tilde{\phi}_1(X) = 0$; let its minimal monic polynomial over $\mathfrak{o}/\mathfrak{p}$ be $\psi_1(X)$. Thus ψ_1 divides $\tilde{\phi}_1$ and, using Lemma 18 again, $\tilde{\phi}_1$ must be a power of ψ_1. But $\deg \psi_1$ divides f_1, the degree of $\mathfrak{O}/\mathfrak{P}$ over $\mathfrak{o}/\mathfrak{p}$, and $\deg \tilde{\phi}_1 = e_1 f_1$; so $\tilde{\phi}_1$ must be an e_1-th power. If $\mathfrak{p} | e_1$ the second coefficient of $\tilde{\phi}_1$ must therefore be 0, whence

$$\alpha \text{ in } \mathfrak{O} \text{ and } \mathfrak{p} | e \quad \Longrightarrow \quad \mathfrak{p} \text{ divides } \mathrm{Tr}_{K_\mathfrak{P}/k_\mathfrak{p}}(\alpha); \qquad (42)$$

using this instead of (41) we deduce that $\mathfrak{P}^e | \mathfrak{d}_{K/k}$. If instead $\mathfrak{p} \nmid e_1$ we can choose α in $\mathfrak{O}$ so that $\tilde{\alpha}$ generates $\mathfrak{O}/\mathfrak{P}$ over $\mathfrak{o}/\mathfrak{p}$ and the second coefficient of ψ_1, which is $-\mathrm{Tr}(\tilde{\alpha})$, is non-zero. But the first of these properties implies that $\tilde{\phi}_1 = \psi_1^{e_1}$, and the second property then shows that $\mathfrak{p}$ does not divide $\mathrm{Tr}_{K_\mathfrak{P}/k_\mathfrak{p}}(\alpha)$; thus $\mathfrak{P}^e$ does not divide $\mathfrak{d}_{K/k}$. $\square$

Corollary *In any extension $K/\mathbf{Q}$ at least one prime p ramifies.*

Proof By the Corollary to Theorem 10, $|d_K| > 1$; hence $\eth_{K/\mathbf{Q}} \neq (1)$ by (37). But note that in general there is no corresponding result for extensions K/k; for more information see §17. □

The lemma and corollary which follow will be useful in §13.

Lemma 25 *Let K_1, K_2 be Galois over k, and write $K = K_1 K_2$. Then K and $K_1 \cap K_2$ are Galois over k,*

$$[K : K_1] = [K_2 : K_1 \cap K_2],$$

and

$$\text{Gal}(K/(K_1 \cap K_2)) \approx \text{Gal}(K_1/(K_1 \cap K_2)) \times \text{Gal}(K_2/(K_1 \cap K_2)).$$

Proof Any element of K is a rational function of elements of K_1 and K_2. Any embedding $K \to \bar{k}$ restricts for $i = 1, 2$ to an embedding $K_i \to \bar{k}$ which is by hypothesis an automorphism of K_i; so any image of an element of K is a rational function of elements of K_1 and K_2, and therefore lies in K. Thus K is normal over k. Similarly an element of $K_1 \cap K_2$ has all its conjugates over k in both K_1 and K_2, and therefore in $K_1 \cap K_2$; so the latter is normal and even Galois over k.

Now choose α in K_2 so that $K_2 = k(\alpha)$, and let $f(X)$ be a minimal monic polynomial for α over $K_1 \cap K_2$. Clearly α generates K over K_1, so K is separable over K_1 and hence over k. Moreover $f(X)$ is irreducible over K_1; for if we had $f(X) = f_1(X)f_2(X)$ in K_1 with the f_i monic, then the coefficients of the f_i would be in K_1 by hypothesis and in K_2 because they are combinations of conjugates of α over k. Hence they would be in $K_1 \cap K_2$ and f would not be irreducible. But now

$$[K : K_1] = \deg(f) = [K_2 : K_1 \cap K_2].$$

In view of what we have already proved, the natural map

$$\text{Gal}(K/(K_1 \cap K_2)) \to \text{Gal}(K_1/(K_1 \cap K_2)) \times \text{Gal}(K_2/(K_1 \cap K_2))$$

is an injection, for if we know the effect of $\sigma : K \to \bar{k}$ on K_1 and K_2 we know it on K. But we already know that both sides have the same degree; so it is an isomorphism. □

Corollary *Let K_1, K_2 be Galois over $\mathbf{Q}$ with coprime discriminants, and write $K = K_1 K_2$. Then*

$$K_1 \cap K_2 = \mathbf{Q}, \quad [K : \mathbf{Q}] = [K_1 : \mathbf{Q}] \cdot [K_2 : \mathbf{Q}], \quad \mathfrak{d}_{K/\mathbf{Q}} = \mathfrak{d}_{K_1/\mathbf{Q}} \mathfrak{d}_{K_2/\mathbf{Q}}$$

and a base for the integers of K over $\mathbf{Q}$ is given by the product of bases for the integers of K_1 and of K_2 over $\mathbf{Q}$.

Proof It follows from (38) that $\mathfrak{d}_{K_1 \cap K_2/\mathbf{Q}}$ divides both $\mathfrak{d}_{K_1/\mathbf{Q}}$ and $\mathfrak{d}_{K_2/\mathbf{Q}}$; since these are coprime by hypothesis, $\mathfrak{d}_{K_1 \cap K_2/\mathbf{Q}} = (1)$. By the Corollary to Theorem 10, this can only happen if $K_1 \cap K_2 = \mathbf{Q}$; this also gives the relation between degrees.

Now suppose that β is in K_2 and also in $\mathfrak{d}_{K/K_1}^{-1}$. If ξ is any integer in K_2 we showed in the proof of the lemma that the conjugates of $\beta\xi$ over K_1 and over $\mathbf{Q}$ are the same; hence $\mathrm{Tr}_{K_2/\mathbf{Q}}(\beta\xi) = \mathrm{Tr}_{K/K_1}(\beta\xi)$ is an integer, and it follows that β is also in $\mathfrak{d}_{K_2/\mathbf{Q}}^{-1}$. Thus $\mathfrak{d}_{K/K_1}$ divides $\mathfrak{d}_{K_2/\mathbf{Q}}$, and it follows from (38) that $\mathfrak{d}_{K/\mathbf{Q}}$ divides $\mathfrak{d}_{K_1/\mathbf{Q}} \mathfrak{d}_{K_2/\mathbf{Q}}$. So $\mathfrak{d}_{K_1/\mathbf{Q}} \mathfrak{d}_{K_2/\mathbf{Q}} = \mathfrak{d}_{K/\mathbf{Q}}$ will follow from the previous sentence combined with $d_{K_1}^{n_2} d_{K_2}^{n_1} = d_K$, where $n_i = [K_i : \mathbf{Q}]$; and this in turn will follow from the assertion about integral bases. Let $\alpha_1, \dots, \alpha_{n_1}$ be a base for the integers of K_1 and $\beta_1, \dots, \beta_{n_2}$ a base for the integers of K_2; then the $\alpha_i \beta_j$ span K as a $\mathbf{Q}$-vector space and are integers, so they generate a subgroup of index m in the $\mathbf{Z}$-module of integers of K. Let $\sigma_1, \dots, \sigma_{n_1}$ be the elements of $\mathrm{Gal}(K/K_2)$; as in the proof of the lemma, their restrictions to K_1 are the elements of $\mathrm{Gal}(K_1/\mathbf{Q})$. Let $\gamma_1, \dots, \gamma_{n_1}$ be elements of K_2 such that $\sum \gamma_i \alpha_i$ is an integer. The equations

$$\gamma_1 \sigma_j \alpha_1 + \cdots + \gamma_{n_1} \sigma_j \alpha_{n_1} = \text{integer} \quad (1 \leqslant j \leq n_1),$$

for the γ_i, have determinant $\sqrt{d_{K_1}}$; thus each $\gamma_i \sqrt{d_{K_1}}$ is an integer, whence m divides some power of d_{K_1}. A similar statement holds for d_{K_2}; and since d_{K_1}, d_{K_2} are coprime, $m = 1$. $\square$

9 Idèles and Adèles

In discussing 'local-to-global' problems it is often necessary to consider several different v-adic fields simultaneously, where each v may be either a finite or an infinite place. The natural language for this is that of **adèles** and **idèles**; we can use this to express some of our previous results in more elegant terms, but its real importance is in the higher reaches of the theory.

At first sight it might seem natural to form the product of all the k_v; this certainly is a topological ring, and it contains all the elements which could interest us. But there are two things wrong with it. It has no satisfactory compactness properties since the individual factors are only locally compact; and most of its elements have not even the most superficial claims to be in the image of the natural map $k \to \prod k_v$, for we know that any α in k is a $\mathfrak{p}$-adic integer for almost all $\mathfrak{p}$. This last remark gives us a strong hint about the set which we ought to be considering, and it turns out that there is only one topology which we can reasonably impose on it.

An **adèle** is an element of the set-theoretic product $\prod k_v$, subject to the condition that if $\alpha = \prod \alpha_v$ is an adèle then α_v is in $\mathfrak{o}_v$ for all but finitely many v. We shall usually denote adèles or idèles by bold Greek letters. They are both examples of what in a more general context is called a **restricted direct product**.

The adèles form a commutative ring V_k under componentwise addition and multiplication. We give V_k a topology by taking as a base for the open sets the $\prod U_v$, where each U_v is open in k_v and $U_v = \mathfrak{o}_v$ for all but finitely many v. It is easy to verify that this does define a topology, that the ring operations are continuous, and that the subspace $\prod \mathfrak{o}_v$ is open and the induced topology on it is just the product topology; hence $\prod \mathfrak{o}_v$ is locally compact and so is V_k.

The **diagonal map** is the map $k \to V_k$ defined by $\alpha \mapsto \prod \alpha$; it enables us to identify k with a subset of V_k. The elements of its image are called the **principal adèles**. This map induces on k the subspace topology; and if we view k and V_k simply as additive groups we can form V_k/k and endow it with the quotient topology.

Lemma 26 *With the conventions above, V_k/k is compact and k has the discrete topology. Moreover $\prod \|\alpha\|_v = 1$ for any $\alpha \neq 0$ in k.*

Proof Theorem 7 shows that given any adèle $\prod \alpha_v$ there exists α in k such that each $\alpha_v - \alpha$ is in $\mathfrak{o}_v$. In other words, every coset of k in V_k meets $\prod' k_v \times \prod'' \mathfrak{o}_\mathfrak{p}$ where the first product is over all infinite places and the second over all finite ones. But $(\prod' k_v)/\mathfrak{o}_k$ is compact, so there is a compact subset $\mathcal{S}$ of $\prod' k_v$ which meets every coset of $\mathfrak{o}_k$. Thus every coset of k in V_k meets $\mathcal{S} \times \prod'' \mathfrak{o}_\mathfrak{p}$, and the latter is compact; hence so is V_k/k.

To prove that k has the discrete topology, it is enough to show that there is a neighbourhood of 0 in V_k which contains no other point of k. But the set defined by

$$\|\alpha_v\|_v < 1 \text{ if } v \text{ is an infinite place,}$$
$$\alpha_v \text{ is in } \mathfrak{o}_v \text{ if } v \text{ is a finite place}$$

is an open set containing 0; and it contains no other element of k by the
Product Formula.   □

The proof that V_k/k is compact, like the results which lead up to it,
does not depend on the Product Formula; so we can now give a new proof
of the latter, which is more highbrow than the one on page 35 but does
explain why such a formula exists. For V_k is locally compact and therefore
possesses an essentially unique Haar measure, which must be the product
of the Haar measures on the factors. Now let α be any non-zero element
of k; multiplication by α defines an automorphism of V_k which maps k
onto itself, so it induces a homeomorphism of V_k/k. Because V_k/k has
finite non-zero measure, this homeomorphism is measure-preserving. But
the action of α on k_v multiplies the Haar measure on k_v by $\|\alpha\|_v$, by the
definition of the normalized valuation; so it multiplies the measure on V_k
by $\prod \|\alpha\|_v$.

Lemma 26 depends crucially on using every place v in forming the product. Let V^w be formed in the same way as V_k but leaving out the factor
corresponding to w; then it is conjectured (but not proved in general) that
k is dense in V^w.

For the argument above, we did not need to normalize the Haar measure
on V_k; but for the following corollary we do. The normalizations we need
are those described in §A1.3.

Corollary *With these normalizations, V_k/k has measure 1.*

Proof As in the proof of the lemma, $V_k/k \approx ((\prod' k_v)/\mathfrak{o}) \times \prod'' \mathfrak{o}_\mathfrak{p}$. The
measure of the second factor on the right is $\prod(\mathrm{Norm}\,\mathfrak{o}_\mathfrak{p})^{-1/2} = |d_k|^{-1/2}$,
and the calculation on page 3 shows that the measure of the first factor is
$|d_k|^{1/2}$.   □

The invertible elements of V_k form a group, called the **idèle group** J_k;
thus an idèle is an element $\alpha = \prod \alpha_v$ such that $\alpha_v \neq 0$ for all v and α_v is a
unit for all but finitely many v. There is a natural map $J_k \to I_k$, where I_k is
the group of ideals; it sends $\alpha = \prod \alpha_v$ to $(\alpha) = \prod \mathfrak{p}^{n_\mathfrak{p}}$ where $\mathfrak{p}^{n_\mathfrak{p}} \|\alpha_\mathfrak{p}$, and it
extends the natural map $k^* \to I_k$. (The infinite places play no part in this
map.) But although J_k is a subset of V_k we must not give it the subspace
topology, for $\xi \mapsto \xi^{-1}$ would not be continuous in that topology. (Take
$k = \mathbf{Q}$ and let $\alpha^{(p)}$ be the adèle with $\alpha_p^{(p)} = p$ and all other components 1;
in the adèlic topology $\alpha^{(p)} \to 1$ as $p \to \infty$, but $(\alpha^{(p)})^{-1}$ does not tend to a
limit.) Instead we give J_k the topology induced by regarding it as a group
of operators on the additive group V_k: that is, a base for the open sets in

J_k is given by the $\prod U_v$ where each U_v is open in k_v^* and $U_v = \mathfrak{o}_v^*$ for all but finitely many v. This topology is strictly finer than the subspace topology, so the inclusion $J_k \to V_k$ and the multiplication map $J_k \times V_k \to V_k$ are continuous; and with it J_k is a locally compact topological group. As with adèles, there is a natural map $k^* \to J_k$ defined by $\alpha \mapsto \prod \alpha$; it is called the **diagonal map**, and enables us to identify k^* with a subset of J_k. The elements of its image are called the **principal idèles**. This map induces on k^* the subspace topology; and we can form J_k/k^* and endow it with the quotient topology.

Lemma 27 *The group k^* is a discrete subgroup of J_k.*

Proof The topology induced on k^* as a subset of J_k is finer than that induced on it as a subset of V_k; and the latter is already the discrete topology, by Lemma 26. $\qquad\square$

The analogue of the first statement in Lemma 26 is much deeper than that lemma, and to state it we need a further definition. The map $J_k \to \mathbf{R}^*$ given by $\alpha = \prod \alpha_v \mapsto \|\alpha\| = \prod \|\alpha_v\|_v$ is well-defined because almost all the factors on the right are equal to 1; and it is a continuous epimorphism. Its kernel J_k^1 contains k^* by the Product Formula. For J_k^1 we no longer have the nuisance of having two distinct induced topologies:

Lemma 28 J_k^1 *is closed both as a subset of J_k and as a subset of V_k, and the two induced topologies on it coincide.*

Proof To prove that J_k^1 is closed in V_k, take any $\prod \alpha_v$ in V_k but not in J_k^1; we shall construct a neighbourhood of it in V_k which does not meet J_k^1. Write $C = \prod \|\alpha_v\|_v$, where the product must either converge or diverge to 0, because it only contains a finite number of terms which exceed 1. Since $C \neq 1$ by hypothesis, there are two cases to consider.

First suppose that $C > 1$. For each infinite place and for those finite primes $\mathfrak{p}$ for which either Norm $\mathfrak{p} \leqslant 2C$ or $\|\alpha_\mathfrak{p}\|_\mathfrak{p} \neq 1$ (of which there can only be finitely many because the product for C converges) let S_v be a small open neighbourhood of α_v in k_v. Now $\prod' \|\alpha_v\|_v = C$ where the product is taken over these v; so we can choose these neighbourhoods so small that if β_v is in S_v for each such v, $\prod' \|\beta_v\|_v < 2C$. For any other v we take $S_v = \mathfrak{o}_v$, so that β_v in S_v implies that either $\|\beta_v\|_v = 1$ or $\|\beta_v\|_v < 1/2C$. If the latter possibility ever happens, then $\prod \|\beta_v\|_v < 1$. Thus $\prod S_v$ is open in the adèlic topology, contains $\prod \alpha_v$ and does not meet J_k^1.

If instead $C < 1$ choose a finite set of places, including all infinite places and all primes with $\|\alpha_\mathfrak{p}\|_\mathfrak{p} > 1$, such that if C' is the product of the $\|\alpha_v\|_v$ taken over these places then $C' < 1$. For each such v let $\mathcal{S}_v$ be a small open neighbourhood of α_v in k_v; these neighbourhoods are to be so small that if β_v is in $\mathcal{S}_v$ for each such v then $\frac{1}{2}(C'+1) > \prod \|\beta_v\|_v$ where the product is over this finite set of v. For any other v take $\mathcal{S}_v = \mathfrak{o}_v$. Thus $\prod \mathcal{S}_v$ is open in the adèlic topology, contains $\prod \alpha_v$ and does not meet J_k^1.

Thus J_k^1 is a closed subset of V_k. It is closed in J_k too, because the idèlic topology on J_k is finer than the restriction of the adèlic topology.

For the last assertion in the lemma it is enough to show that any J_k-open subset of J_k^1 is V_k-open, the converse being trivial. Now let $\mathcal{S} = \prod \mathcal{S}_v$ be any basic J_k-open set; we need to find a V_k-open set $\mathcal{S}'$ such that $\mathcal{S} \cap J_k^1 = \mathcal{S}' \cap J_k^1$. By writing $\mathcal{S}$ as a union of smaller basic open sets if necessary, we can assume that each $\mathcal{S}_v$ is bounded; since for all but finitely many v we have $\mathcal{S}_v = \mathfrak{o}_v^*$ and therefore $\|\alpha_v\|_v = 1$ for all α_v in $\mathcal{S}_v$, there is a constant C such that $\prod \|\alpha_v\|_v < C$ for all $\prod \alpha_v$ in $\mathcal{S}$. Now write

$$\mathcal{S}'_v = \begin{cases} \mathfrak{o}_\mathfrak{p} \text{ if } \mathfrak{p} \text{ is finite, } \mathcal{S}_\mathfrak{p} = \mathfrak{o}_\mathfrak{p}^* \text{ and Norm} \, \mathfrak{p} \geqslant 2C, \\ \mathcal{S}_v \text{ otherwise.} \end{cases}$$

Since the first of these happens for all but finitely many $\mathfrak{p}$, $\mathcal{S}' = \prod \mathcal{S}_v$ is open in V_k; and $\mathcal{S}' \cap J_k^1 = \mathcal{S} \cap J_k^1$ as in the first part of the proof. $\qquad\square$

Theorem 22 J_k^1/k^* *is compact.*

Proof As before, denote by I_k the multiplicative group of non-zero fractional ideals of k, and endow it with the discrete topology. Because $I_k/\{\text{image of } k^*\}$ is finite, there is a finite group of translates of $\prod \mathfrak{o}_v^*$ whose union meets every coset of k^* in J_k^1. By Theorem 11 there is a closed bounded set $\mathcal{S}$ in $\prod' k_v^*$, where the product is taken over the infinite places, such that any coset of k^* in J_k^1 which meets $\prod' k_v^* \times \prod'' \mathfrak{o}_\mathfrak{p}^*$ also meets $\mathcal{S} \times \prod'' \mathfrak{o}_\mathfrak{p}^*$. But this last set is the product of compact spaces and is therefore compact. $\qquad\square$

To prove Theorem 22 we used the finiteness of the ideal class group and the structure of the group of units. Conversely, from an independent proof of Theorem 22 we can immediately deduce these two results — which are the key structural theorems of the elementary theory. For such a proof, see Chapter II of [CF].

We specified the measure on J_k in §A1.3, and the measure on J_k^1 now follows from the exact sequence (algebraic and topological)

$$0 \to J_k^1 \to J_k \to \mathbf{R}^* \to 0.$$

In this case we can be very explicit, though uncanonical, because the exact sequence splits. Let u be a fixed one of the infinite primes, and let α be any idèle. There is a unique β in J_k^1 which is the same as α in all but the u-th component and for which α_u/β_u is real and positive; and the map $\phi : \alpha \mapsto \beta$ is a continuous homomorphism. Given a set $\mathcal{S}^1$ in J_k^1, let $\mathcal{S}$ be that part of its inverse image under ϕ which satisfies

$$0 \leqslant \log \|\alpha\| = \log \prod \|\alpha_v\|_v \leq 1$$

and define the measure of $\mathcal{S}^1$ to be the measure of $\mathcal{S}$. It is easy to see that this gives a Haar measure on J_k^1, which does not depend on the choice of u.

Theorem 23 *If the measure on J_k is the product of the measures above on the individual k_v^* then the measure of J_k^1/k^* is $2^{r_1}(2\pi)^{r_2} hR/w$ where w is the number of roots of unity in k^*.*

Proof The map $J_k \to I_k$ induces an epimorphism from J^1 to the ideal class group; if we denote the kernel of this epimorphism by $\mathcal{S}^1$ then it is enough to find the measure of $\mathcal{S}^1/k^*$ and multiply by h. Each coset of k^* in $\mathcal{S}^1$ contains elements of the set $\mathcal{S}^\sharp$ consisting of the idèles in $\mathcal{S}^1$ whose components at each finite prime $\mathfrak{p}$ are in $\mathfrak{o}_\mathfrak{p}^*$; and these elements are determined up to an element of $\mathfrak{o}^*$. Denote by $\mathcal{S}^\flat$ the elements of the finite product $\prod' \alpha_v$ such that $\prod' \|\alpha_v\|_v = 1$, where the products are taken over the infinite places. There is a natural measure on $\mathcal{S}^\flat$ induced by the measures chosen on $\mathbf{R}^*$ and $\mathbf{C}^*$, and the measures of $\mathcal{S}^1/k^*$, of $\mathcal{S}^\sharp/\mathfrak{o}_k^*$ and of $\mathcal{S}^\flat/\mathfrak{o}_k^*$ are all equal since $\mathcal{S}^\sharp = \mathcal{S}^\flat \times \prod \mathfrak{o}_\mathfrak{p}^*$. Now let $\eta_1, \ldots, \eta_{r_1+r_2-1}$ be a base for the units modulo roots of unity; given any $\prod' \alpha_v$ in $\mathcal{S}^\flat$ we can define real numbers $x_1, \ldots, x_{r_1+r_2-1}$ by the equations

$$\log \|\alpha_v\|_v = x_1 \log \|\eta_1\|_v + \cdots + x_{r_1+r_2-1} \log \|\eta_{r_1+r_2-1}\|_v$$

for each Archimedean v. Each coset of $\mathfrak{o}^*$ in $\mathcal{S}^\flat$ has just w members in the region given by $0 \leqslant x_i < 1$. But if v is real, going from $d\alpha_v/\|\alpha_v\|_v$ to $d(\log \|\alpha_v\|_v)$ gives a factor 2 to take account of the sign of α_v; and if v is complex, $dz \wedge d\bar{z}/|z|^2 = d(r^2) \wedge d\theta/r^2$ in polar coordinates, and integrating over θ gives a factor 2π. The measure of $\mathcal{S}^\flat/\mathfrak{o}_k^*$ is therefore

$$2^{r_1}(2\pi)^{r_2} w^{-1} \int \cdots \int \prod'' d(\log \|\alpha_v\|_v)$$

where the product is taken over all but one of the Archimedean valuations and the multiple integral is taken over the region given by $0 \leqslant x_i < 1$

for each i. Going from the $\log \|\alpha_v\|_v$ to the x_i gives a further factor $|R|$. Combining these factors, we obtain the theorem.　　□

3

Special fields

Most of this chapter illustrates the general theory in Chapters 1 and 2 by means of applications to particular kinds of field. However, many of the results in §12 and §13.1 depend on analytic properties of the zeta function which are not stated until §14 and not proved until §15. The standard calculations on $\mathbf{Q}(\sqrt[n]{1})$ in §13 are also used to prove a particular case of Fermat's Last Theorem, due to Kummer.

10 Quadratic fields

The quadratic fields are just the fields $k = \mathbf{Q}(\sqrt{m})$ where $m \neq 1$ is a square-free integer. Since $1, \sqrt{m}$ are integers which form a base for k as a $\mathbf{Q}$-vector space and $\Delta^2(1, \sqrt{m}) = 4m$ is not divisible by any square other than 4, the $\mathbf{Z}$-module spanned by $1, \sqrt{m}$ either is $\mathfrak{o}$ or has index 2 in $\mathfrak{o}$. Since $\frac{1}{2}$ and $\frac{1}{2}\sqrt{m}$ are clearly not integers, the latter case happens if and only if $\alpha = \frac{1}{2}(1 + \sqrt{m})$ is an integer. The minimal equation for α over $\mathbf{Q}$ is $X^2 - X - \frac{1}{4}(m - 1) = 0$, so α is an integer if and only if $m \equiv 1 \bmod 4$. Thus

$$d = \begin{cases} m & \text{if} \quad m \equiv 1 \bmod 4, \\ 4m & \text{if} \quad m \equiv 2 \text{ or } 3 \bmod 4. \end{cases}$$

Both cases are covered by the statement that the integers are those numbers of the form $\frac{1}{2}(x + y\sqrt{d})$ for which x, y and $\frac{1}{4}(x^2 - dy^2)$ are in $\mathbf{Z}$. The primes which ramify are those which divide d, and ramification can only mean $(p) = \mathfrak{p}^2$ with $f = 1$. Any other prime either splits (that is, $(p) = \mathfrak{p}'\mathfrak{p}''$ with $f = 1$ for each factor) or remains prime (with $f = 2$). To test which, we use Theorem 19 and Lemma 19; for $X^2 - m$ factorizes in $\mathbf{Q}_p$ if and only if

55

$X^2 - m = 0$ has a root in $\mathbf{Q}_p$. Thus when 2 does not ramify

(2) splits in k if $m \equiv 1 \bmod 8$,

(2) remains prime in k if $m \equiv 5 \bmod 8$;

and if p is an odd prime which does not divide m,

(p) splits in k if and only if $\left(\tfrac{m}{p}\right) = 1$

where the latter bracket is the quadratic residue symbol.

If $m > 0$ then k is real and the only roots of unity which it contains are ± 1. If $m < 0$ all units in k are roots of unity, and they correspond to the solutions of $x^2 - dy^2 = 4$ in $\mathbf{Z}$. It is now easy to see that k contains six roots of unity if $d = -3$, four if $d = -4$ and two otherwise.

There is a close relation between ideal classes in a quadratic field k and classes of binary quadratic forms

$$aX_1^2 + bX_1X_2 + cX_2^2 \quad \text{with} \quad b^2 - 4ac = d \tag{43}$$

under the unimodular group. For let $\mathfrak{a}$ be an non-zero integral ideal of k and let α_1, α_2 be a base for $\mathfrak{a}$ as a $\mathbf{Z}$-module. Any α in $\mathfrak{a}$ has the form $\alpha = x_1\alpha_1 + x_2\alpha_2$ with x_1, x_2 in $\mathbf{Z}$, which implies

$$\text{norm}\,\alpha = (x_1\alpha_1 + x_2\alpha_2)(x_1\sigma\alpha_1 + x_2\sigma\alpha_2)$$

where σ is the non-trivial element of $\mathrm{Gal}(k/\mathbf{Q})$. The right hand side is a quadratic form in x_1, x_2 all of whose coefficients are rational integers divisible by $\mathfrak{a} \cdot \sigma\mathfrak{a} = (\text{Norm}\,\mathfrak{a})$, and its discriminant is

$$\begin{vmatrix} \alpha_1 & \alpha_2 \\ \sigma\alpha_1 & \sigma\alpha_2 \end{vmatrix}^2 = (\text{Norm}\,\mathfrak{a})^2 d_k.$$

Dividing by Norm $\mathfrak{a}$ we get a quadratic form (43). We can change the base for $\mathfrak{a}$ as a $\mathbf{Z}$-module by means of an integral unimodular transformation on x_1, x_2. If we start from another ideal in the same class — say $\mathfrak{a}' = (\beta)\mathfrak{a}$ for some β in k^* — then we can take $\beta\alpha_1, \beta\alpha_2$ as a base for $\mathfrak{a}'$; this yields the same quadratic form as before. Conversely if we write

$$aX_1^2 + bX_1X_2 + cX_2^2 = a(X_1 + \lambda X_2)(X_1 + \sigma\lambda X_2)$$

a straightforward calculation shows that $a, a\lambda$ are a base for an ideal of k and this ideal gives rise to the quadratic form which we started from. However, the quadratic form has two linear factors and therefore gives rise to two ideal classes, which are in general distinct.

When $m < 0$ there is a simple way round this: we fix an embedding $k \to \mathbf{C}$ and require the base α_1, α_2 to satisfy $\Im(\alpha_2/\alpha_1) > 0$. This imposes an ordering on any base for $\mathfrak{a}$ as a **Z**-module, and restricts us to integral transformations on x_1, x_2 with determinant $+1$. Now there is a one-one correspondence between ideal classes in k and equivalence classes of binary quadratic forms (43). This gives the easiest way of computing class numbers when $m < 0$. For after a suitable linear transformation we can assume that a is the least value taken by the quadratic form (43) and that, subject to this, c is as small as possible; in this case we say that (43) is in **reduced form**. This gives $c \geqslant a$ and $|b| \leqslant a$, whence $a \leqslant \sqrt{-d/3}$. Conversely, if these inequalities hold it follows easily that a and c have the properties in the previous sentences. We can obtain a unique representative of the equivalence class of quadratic forms by writing $X_1 + X_2$ for X_1 if $b = -a$, or $X_2, -X_1$ for X_1, X_2 if $c = a$ and $b < 0$; in this way we can require $a \geqslant b > -a$, $c \geqslant a$ and $b \geqslant 0$ if $c = a$. If $\beta = \alpha_2/\alpha_1$ these conditions are equivalent to $\frac{1}{2} \geqslant \Im\beta > -\frac{1}{2}$, $|\beta| \geqslant 1$ and $\Im\beta \geqslant 0$ if $|\beta| = 1$. The reader who knows about elliptic modular functions will recognize the fundamental domain of the modular group.

To compute the ideal class number, we let b run through all values with $|b| \leqslant \sqrt{-d/3}$, where b is odd or even according as d is; for each such b we factorize $ac = \frac{1}{4}(b^2 - d)$ and list the triplets (a, b, c) with $a \geqslant b > -a$, $c \geqslant a$ and $b \geqslant 0$ if $a = c$. There are h of these triplets, where h is the class number of k. The Brauer-Siegel Theorem (76) says that $\log h \sim \frac{1}{2} \log |d|$ as $d \to -\infty$, but the known effective results are much weaker. There is a law of composition of quadratic forms (43) which corresponds to multiplication of ideal classes, but for computational purposes there is little to be gained by using it. There is a remarkable conjecture, supported by both theoretical arguments and numerical evidence, that the odd order part of the ideal class group is cyclic for 97·7% of all values of $m < 0$. (See [Co], §5.10.)

Theorem 24 *Let $m < 0$ and suppose that t distinct primes divide d. Then there are exactly 2^{t-1} elements of order 1 or 2 in the ideal class group of k.*

Proof Because $\mathfrak{a} \cdot \sigma\mathfrak{a} = (\text{Norm } \mathfrak{a})$ is principal, we need to count the number of ideal classes fixed by σ. But if α_1, α_2 are a base of $\mathfrak{a}$ with $\Im(\alpha_2/\alpha_1) > 0$ then $\sigma\alpha_1, -\sigma\alpha_2$ are a base of $\sigma\mathfrak{a}$ with $\Im(-\sigma\alpha_2/\sigma\alpha_1) > 0$. Suppose that the reduced binary quadratic form corresponding to $\mathfrak{a}$ is $aX_1^2 + bX_1X_2 + cX_2^2$; then $\sigma\mathfrak{a}$ corresponds to $aX_1^2 - bX_1X_2 + cX_2^2$ and this is also reduced. But $\mathfrak{a}$ and $\sigma\mathfrak{a}$ are in the same ideal class if and only if these forms are equivalent, and it is easy to see that this happens precisely for forms of the shape

$$aX_1^2 + cX_2^2, \quad aX_1^2 \pm aX_1X_2 + cX_2^2 \quad \text{or} \quad aX_1^2 + bX_1X_2 + aX_2^2.$$

In each of these cases the ideals $\mathfrak{a} = (a, a\beta)$ and $\sigma\mathfrak{a} = (a, -a\sigma\beta)$, where $a\beta^2 + b\beta + c = 0$, are not merely in the same class but equal. Hence each ideal class in k of order 1 or 2 contains at least one ideal which is a product of ramified primes and rational primes, and clearly we can ignore the latter. Conversely, the square of any ideal which is a product of ramified primes is principal.

It remains to discover which of the 2^t ideals generated in this way are themselves principal. Each of the latter corresponds to two pairs of integers x, y and $-x, -y$ such that $\frac{1}{4}(x^2 - dy^2)$ is square-free and divides $-d$. This implies $x^2 - dy^2 \leqslant -4d$, so that $|y| \leqslant 2$. Apart from the obvious solutions $(\pm 2, 0)$ and $(0, \pm 2)$, the latter requiring $4 \nmid d$, we need only consider $y = \pm 1$. Now $x^2 - dy^2 = -4d/r$ for some r requires $1 \leqslant r \leqslant 4$; since $d = -2$ is not allowed, the only solutions are given by $x = 0$ when $4 | d$, $x = \pm 1$ when $d = -3$, $x = 2$ when $d = -4$, and $x = \pm 3$ when $d = -3$. Thus for every value of d we obtain just two principal ideals of this kind: these are $(1 + \sqrt{-1})$ and (1) when $m = -1$, and $(\sqrt{m})$ and (1) in all other cases. $\square$

When $m > 0$ the situation is more complicated. By Theorem 11 the group of units is the product of $\{\pm 1\}$ and an infinite cyclic group; any of the four units which generates the latter is called a **fundamental unit**. There are various ways of defining a reduced form (43), the simplest being to require $|c| \geqslant |a| \geqslant |b|$; for given d there are only finitely many reduced forms, and they can be listed because $a^2 < \frac{1}{4}d$. But each equivalence class will contain a number of reduced forms, this number being large if the fundamental unit is large, which it usually is. Moreover, one wishes to find both the class number and the group of units. For hand calculation, the better way to proceed is illustrated in Examples 4 to 6 below. Here we take advantage of the estimates quoted on page 20, whose proof is sketched in the exercise at the end of this section. The analogue of Theorem 24 is that there are 2^{t-1} elements of order 1 or 2 in the ideal class group if the fundamental unit has norm -1, and 2^{t-2} if it has norm $+1$, but the proof is considerably more complicated than that of Theorem 24. Subject to this, the class number is usually small; for example it is conjectured that the class number is not divisible by any odd prime in $75 \cdot 4\%$ of all cases. But even the assertion that $h = 1$ infinitely often is only a long-standing conjecture.

Example 1 $m = -14$. Now $d = -56$ and the reduced quadratic forms (43) are

$$X_1^2 + 14X_2^2, \quad 2X_1^2 + 7X_2^2, \quad 3X_1^2 \pm 2X_1X_2 + 5X_2^2.$$

Thus $h = 4$. The ideal class group must be cyclic, by Theorem 24; and this example shows that we cannot strengthen the conclusion of that theorem to $2^{t-1} \| h$.

Example 2 $m = -35$. Now $d = -35$ and if we take account of equivalences the reduced quadratic forms are

$$X_1^2 + X_1 X_2 + 9X_2^2, \quad 3X_1^2 + X_1 X_2 + 3X_2^2,$$

so that $h = 2$. The odd prime p splits if and only if $\left(\frac{-35}{p} \right) = 1$; such a p is a product of principal ideals if $p = X_1^2 + X_1 X_2 + 9X_2^2$ is soluble in integers, and of non-principal ideals if $p = 3X_1^2 + X_1 X_2 + 3X_2^2$ is soluble in integers. (Just one of these must happen.) The former requires $p \equiv \pm 1 \bmod 5$ and the latter requires $p \equiv \pm 2 \bmod 5$. But to have so simple a rule as this depends on having $h = 2^{t-1}$ in the notation of Theorem 24.

Example 3 $m = -131$. Now $d = -131$ and if we take account of equivalences the reduced quadratic forms are

$$X_1^2 + X_1 X_2 + 33X_2^2, \quad 3X_1^2 \pm X_1 X_2 + 11X_2^2, \quad 5X_1^2 \pm 3X_1 X_2 + 7X_2^2,$$

so that $h = 5$. Here (2) is prime, $(3) = \mathfrak{p}_3' \mathfrak{p}_3''$ and $(5) = \mathfrak{p}_5' \mathfrak{p}_5''$, where we can name the factors so that

$$\sqrt{-131} \equiv 1 \bmod \mathfrak{p}_3' \text{ and } \sqrt{-131} \equiv 2 \bmod \mathfrak{p}_3'',$$
$$\sqrt{-131} \equiv 2 \bmod \mathfrak{p}_5' \text{ and } \sqrt{-131} \equiv 3 \bmod \mathfrak{p}_5''.$$

There is no principal ideal with Norm 3, 5 or 15 because the first quadratic form does not represent any of these numbers; so the four ideals we have produced through factorization must between them represent the four non-principal ideal classes. Hence there must be a principal ideal of Norm 45, and one such is $(\frac{1}{2}(7 + \sqrt{-131}))$. This is not divisible by $\mathfrak{p}_3'$ or $\mathfrak{p}_5'$, so it must factor as $\mathfrak{p}_3''^2 \mathfrak{p}_5''$, whence $\mathfrak{p}_5''$ is in the class of $\mathfrak{p}_3''^3$. Similarly $(7) = \mathfrak{p}_7' \mathfrak{p}_7''$ where $\sqrt{-131} \equiv 3 \bmod \mathfrak{p}_7'$ and $\sqrt{-131} \equiv 4 \bmod \mathfrak{p}_7''$; and consideration of $(\frac{1}{2}(3 + \sqrt{-131}))$ whose Norm is 35 shows that $\mathfrak{p}_7'$ is in the same ideal class as $\mathfrak{p}_5'$.

Example 4 $m = 10$. Now $d = 40$ and $\sqrt{40/5} < 3$, so any ideal class contains an integral ideal of Norm at most 2 and we need only look at how 2 factorizes. We know that $(2) = \mathfrak{p}_2^2$; moreover $\mathfrak{p}_2$ is not principal because $X_1^2 - 10X_2^2 = \pm 2$ is insoluble in $\mathbf{Z}_5$ and hence in $\mathbf{Z}$. Thus $h = 2$. By inspection $3 + \sqrt{10}$ is a unit, and it is the fundamental unit because it gives the least non-trivial solution of $X_1^2 - 10X_2^2 = \pm 1$.

Example 5 $m = 229$. This has been chosen because it is the first case when h is not a power of 2. Now $d = 229$ and $\sqrt{229/8} < 6$, so using the strong result quoted at the end of §3 any ideal class contains an integral ideal of Norm at most 5. Thus we need only consider the factorization of 2, 3 and 5. Here (2) is prime, $(3) = \mathfrak{p}_3'\mathfrak{p}_3''$ and $(5) = \mathfrak{p}_5'\mathfrak{p}_5''$. But $\frac{1}{2}(13 + \sqrt{229})$ has norm -15, so after renumbering if necessary it must be equal to $\mathfrak{p}_3'\mathfrak{p}_5'$; and $\mathfrak{p}_5'$ is in the same class as $\mathfrak{p}_3''$. By inspection $\frac{1}{2}(15 + \sqrt{229})$ is a unit; it is actually a fundamental unit, but we shall not need this fact. To prove that $h = 3$ it is enough to show that $\mathfrak{p}_3'^2$ is not principal — that is, that all integer solutions of $X_1^2 + X_1X_2 - 57X_2^2 = \pm 9$ have X_1, X_2 both divisible by 3 and thus give elements of k in the ideal (3). This time congruence arguments do not help. But if the equation has a solution x_1, x_2, we can multiply $(x_1 + \frac{1}{2}(1 + \sqrt{229})x_2)$ by a power of the known unit $\frac{1}{2}(15 + \sqrt{229})$ to ensure that

$$3 \leqslant |x_1 + \tfrac{1}{2}(1 + \sqrt{229})x_2| < 3 \times \tfrac{1}{2}(15 + \sqrt{229});$$

this implies also

$$3 \times \tfrac{1}{2}(\sqrt{229} - 15) < |x_1 + \tfrac{1}{2}(1 - \sqrt{229})x_2| \leqslant 3.$$

This process does not affect whether x_1, x_2 are both divisible by 3. These inequalities define a bounded search region, and we find that within it there are no pairs x_1, x_2 of the kind we are looking for. Of course, we can make this process more efficient; but that does not change the underlying idea.

Example 6 $m = 73$. This is a case where the fundamental unit is large enough for one to need an efficient process for finding it. Now $d = 73$ and $\sqrt{73/8} < 4$, so using the strong result quoted at the end of §3 any ideal class contains an integral ideal of Norm at most 3. But $(2) = \mathfrak{p}_2'\mathfrak{p}_2''$ and $(3) = \mathfrak{p}_3'\mathfrak{p}_3''$, so after renaming we can require that $(\frac{1}{2}(7 + \sqrt{73})) = \mathfrak{p}_2'\mathfrak{p}_3'$. By considering Norms we obtain

$$
\begin{array}{ll}
(\tfrac{1}{2}(1 + \sqrt{73})) = \mathfrak{p}_2''\mathfrak{p}_3'^2, & \qquad (\tfrac{1}{2}(3 + \sqrt{73})) = \mathfrak{p}_2'^4, \\
(\tfrac{1}{2}(5 + \sqrt{73})) = \mathfrak{p}_2''^2\mathfrak{p}_3', & \qquad (\tfrac{1}{2}(7 + \sqrt{73})) = \mathfrak{p}_2'\mathfrak{p}_3', \\
(\tfrac{1}{2}(9 + \sqrt{73})) = \mathfrak{p}_2'', & \qquad (\tfrac{1}{2}(11 + \sqrt{73})) = \mathfrak{p}_2'^2\mathfrak{p}_3''.
\end{array}
$$

These equations are enough to show that $\mathfrak{p}_2', \mathfrak{p}_2'', \mathfrak{p}_3', \mathfrak{p}_3''$ are all principal, so $h = 1$. Moreover

$$\frac{(\tfrac{1}{2}(7 + \sqrt{73}))(\tfrac{1}{2}(9 + \sqrt{73}))^3(\tfrac{1}{2}(11 + \sqrt{73}))}{24}$$

has trivial divisor, so that it is a unit. It is equal to $1068 + 125\sqrt{73}$ and this is actually a fundamental unit. It might appear from the factorizations above that one could form smaller combinations with trivial divisor, but there is a pitfall to beware of. For example

$$\frac{(\frac{1}{2}(5+\sqrt{73}))(\frac{1}{2}(7+\sqrt{73}))(\frac{1}{2}(9-\sqrt{73}))}{12}$$

has trivial divisor; but it is just a complicated way of expressing 1.

Lemma 29 *Let p be an odd prime and ρ a primitive p-th root of unity. Let $S = \sum \left(\frac{r}{p}\right)\rho^r$ where the bracket is the quadratic residue symbol and the sum is over the residue classes prime to p; then $S^2 = \left(\frac{-1}{p}\right)p$.*

Proof We obtain $S^2 = \sum\sum \left(\frac{rs}{p}\right)\rho^{r+s} = \sum_t \left(\frac{t}{p}\right)\sum_r \rho^{r(1+t)}$ on writing $s = rt$. The inner sum is p if $t = -1$, and 0 otherwise. □

Corollary *Any quadratic field is cyclotomic — that is, it is a subfield of the field of all roots of unity.*

Proof After the lemma, it is enough to prove the result for $\mathbf{Q}(\sqrt{-1})$ and $\mathbf{Q}(\sqrt{2})$, for any quadratic field $\mathbf{Q}(\sqrt{m})$ lies in the least field containing $\sqrt{-1}$, $\sqrt{2}$ and the $\sqrt{\pm p}$ for the odd primes p dividing m. But $\sqrt{-1}$ is a fourth root of unity and $(1 + \sqrt{-1})/\sqrt{2}$ is an eighth root of unity. The corollary is also a special case of the Kronecker-Weber Theorem, for which see §20. □

Theorem 25 (Quadratic Reciprocity) *If p, q are distinct odd primes, then*

$$\left(\frac{p}{q}\right)\left(\frac{q}{p}\right) = (-1)^{(p-1)(q-1)/4}. \tag{44}$$

Proof Since $\mathbf{F}_p^*$ is cyclic, $\left(\frac{r}{p}\right) \equiv r^{(p-1)/2} \bmod p$. If S is as in Lemma 29 then

$$S^{q-1} = (-1)^{(p-1)(q-1)/4}p^{(q-1)/2} \equiv (-1)^{(p-1)(q-1)/4}\left(\frac{p}{q}\right) \bmod q.$$

On the other hand, if we write $s = rq$,

$$S^q \equiv \sum \left(\frac{r}{p}\right)\rho^{rq} = \left(\frac{q}{p}\right)\sum\left(\frac{s}{p}\right)\rho^s = \left(\frac{q}{p}\right)S \bmod q;$$

dividing by S, which is prime to q, and combining with the previous equation we obtain (44). □

To obtain the auxiliary law $\left(\frac{2}{p}\right) = (-1)^{(p^2-1)/8}$, consider $T = (1+i)^p$ where $i = \sqrt{-1}$. For on the one hand $T \equiv 1 + i^p \bmod p$; on the other hand

$$T = (1+i)(2i)^{(p-1)/2} \equiv (1+i)i^{(p-1)/2}\left(\frac{2}{p}\right) \bmod p.$$

Now consider separately the four possible values of $p \bmod 8$.

Exercise Let $f = aX^2 + bXY + cY^2$ with $d = b^2 - 4ac \neq 0$ and a, b, c real have $\min |f(x, y)| = M > 0$ where x, y run through all pairs of rational integers not both zero. If the minimum M is attained, show that by an integral unimodular change of variables, and replacing f by $-f$ if necessary, one can take $a = M$, $0 \leqslant b \leqslant M$.

(i) If $d < 0$ show that M is attained and $d \geqslant 3M^2$.

(ii) If $d > 0$ and M is attained, then split cases by considering $f(1, 1)$. Show that $f(1, 1) \geqslant M$ implies that $f = M(X^2 + XY - Y^2)$ and $d = 5M^2$. If $f(1, 1) \leqslant -M$ then $f = M(X^2 - 2Y^2)$ and $d = 8M^2$ if $f(3, -2) \geqslant M$, while $d \geqslant \frac{221}{25}M^2$ if $f(3, -2) \leqslant -M$.

In (ii), can the argument be pushed further? What happens if M is not attained?

11 Pure cubic fields

These are fields $\mathbf{Q}(\sqrt[3]{m})$, where we can take $m = m_1 m_2^2$ with m_1, m_2 coprime, square-free and positive. Write

$$\alpha_1 = \sqrt[3]{m_1 m_2^2}, \quad \alpha_2 = \sqrt[3]{m_1^2 m_2}$$

so that $1, \alpha_1, \alpha_2$ span k as a $\mathbf{Q}$-vector space. We have

$$\Delta^2(1, \alpha_1, \alpha_2) = -27 m_1^2 m_2^2.$$

Theorem 26 *In the notation above, $d = -27 m_1^2 m_2^2$ and $1, \alpha_1, \alpha_2$ are a base for $\mathfrak{o}$ unless $m_1 \equiv \pm m_2 \bmod 9$. In the latter case $\frac{1}{3}(1 + m_1 \alpha_1 + m_2 \alpha_2)$ is also in $\mathfrak{o}$ and $d = -3 m_1^2 m_2^2$.*

Proof If for example $p \neq 3$ is a prime factor of m_1 then $(p, \alpha_1)^3 = (p)$, so that p ramifies; since $p^2 \| \Delta^2$ we must have $p^2 \| d$. If $3 | m_1$ a similar calculation shows that $(3) = \mathfrak{p}_3^3$ with $\mathfrak{p}_3 \| \alpha_1$ and $\mathfrak{p}_3^2 \| \alpha_2$. Hence if c_0, c_1, c_2 are in $\mathbf{Q}$ the exact powers of $\mathfrak{p}_3$ which divide the non-zero summands in $\alpha = c_0 + c_1 \alpha_1 + c_2 \alpha_2$ are all different, so that α can only be an integer if c_0, c_1, c_2 are 3-adic integers. It follows that $1, \alpha_1, \alpha_2$ are a base for $\mathfrak{o}$ if $3 | m_1$ or similarly if $3 | m_2$. If instead $3 \nmid m_1 m_2$ then d contains an odd power of 3 and therefore 3 ramifies. Theorem 11 allows two possibilities:

$$(3) = \mathfrak{p}_3' \mathfrak{p}_3''^2 \text{ and } d = -3 m_1^2 m_2^2, \quad \text{or} \quad (3) = \mathfrak{p}_3^3 \text{ and } d = -27 m_1^2 m_2^2.$$

By Theorem 19 the first case happens if and only if $X^3 - m = 0$ is soluble in $\mathbf{Q}_3$, and by Lemma 19 this happens if and only if $m \equiv \pm 1 \bmod 9$. This is the same as $m_1 \equiv \pm m_2 \bmod 9$. To find a base for the integers in this case, let $\mu = \sqrt[3]{m}$ in $\mathbf{Q}_3$; then $(\alpha_1 - \mu)(\alpha_1^2 + \mu\alpha_1 + \mu^2) = 0$, so Theorem 19 implies that $m_2\alpha_2 + \mu\alpha_1 + \mu^2 = 0$ in $k_\mathfrak{p}$ where $\mathfrak{p} = \mathfrak{p}_3''$. Since $\mu \equiv m_1 \bmod 3$ it follows easily that $1 + m_1\alpha_1 + m_2\alpha_2 \equiv 0 \bmod 3$. $\qquad\square$

Now suppose that $p \nmid 3m_1 m_2$, so that p is not ramified. Since $\mathbf{F}_p^*$ is cyclic of order $p - 1$, the equation $X^3 - 1 = 0$ has one root in $\mathbf{F}_p$ if $p \equiv 2 \bmod 3$ and three if $p \equiv 1 \bmod 3$. Thus if $p \equiv 2 \bmod 3$ then $X^3 - \widetilde{m} = 0$ has one root in $\mathbf{F}_p$ and that root has multiplicity 1; so it follows from Theorem 19 that $(p) = \mathfrak{p}'\mathfrak{p}''$ where $\mathrm{Norm}\,\mathfrak{p}' = p$ and $\mathrm{Norm}\,\mathfrak{p}'' = p^2$. If $p \equiv 1 \bmod 3$ then $x^3 - \widetilde{m} = 0$ has three roots in $\mathbf{F}_p$ or none; and the theoretical criteria for distinguishing between these two possibilities are more complicated than direct calculation. In the first case $(p) = \mathfrak{p}'\mathfrak{p}''\mathfrak{p}'''$; in the second case (p) is a prime ideal in k.

Example $m = 12$. Now $1, \alpha_1 = \sqrt[3]{12}, \alpha_2 = \sqrt[3]{18}$ are a base for $\mathfrak{o}$ and

$$\mathrm{norm}\,(c_0 + c_1\alpha_1 + c_2\alpha_2) = f(c_0, c_1, c_2) = c_0^3 + 12c_1^3 + 18c_2^3 - 18c_0c_1c_2.$$

Since $\sqrt{-d/23} < 7$ we need only look at the factorization of 2, 3 and 5. We have already seen that $(2) = \mathfrak{p}_2^3$, $(3) = \mathfrak{p}_3^3$ and $(5) = \mathfrak{p}_5'\mathfrak{p}_5''$ where $\mathrm{Norm}\,\mathfrak{p}_5' = 5$ and $\mathrm{Norm}\,\mathfrak{p}_5'' = 5^2$. But

$$
\begin{aligned}
f(2,1,1) &= 2 && \text{so that} && (2 + \alpha_1 + \alpha_2) = \mathfrak{p}_2, \\
f(3,1,1) &= 3 && \text{so that} && (3 + \alpha_1 + \alpha_2) = \mathfrak{p}_3, \\
f(5,2,2) &= 5 && \text{so that} && (5 + 2\alpha_1 + 2\alpha_2) = \mathfrak{p}_5', \\
f(0,-1,1) &= 6 && \text{so that} && (\alpha_1 - \alpha_2) = \mathfrak{p}_2\mathfrak{p}_3.
\end{aligned}
$$

It follows from the first three of these that all ideals of Norm less than 7 are principal; so $h = 1$. Moreover

$$\frac{(2 + \alpha_1 + \alpha_2)(3 + \alpha_1 + \alpha_2)}{\alpha_2 - \alpha_1} = 55 + 24\alpha_1 + 21\alpha_2$$

is a unit — and actually the fundamental unit.

12 Biquadratic fields

Let K be the biquadratic field $\mathbf{Q}(\sqrt{a_1}, \sqrt{a_2})$, where none of a_1, a_2 and a_1a_2 are squares, and write $a_3 = a_1a_2/m^2$ where m^2 is the largest square which divides a_1a_2. Then K is normal over $\mathbf{Q}$ with Galois group $C_2 \times C_2$, and K has three intermediate fields $k_i = \mathbf{Q}(\sqrt{a_i})$. Throughout this section

d, h, R and w will be the values for K, and the values for k_i will be denoted by the subscript i. Neglect for the moment primes which ramify in $K/\mathbf{Q}$; then $e = 1$ and the splitting group Z is cyclic of order f. We have two possibilities:

(i) $f = 1$, so that p splits completely in K and in each k_i;
(ii) $f = 2$, so that p splits in one k_i and remains prime in the other two.

Thus up to finitely many factors of the form $(1 - p^{-fs})^{\pm 1}$ arising from the ramified primes,

$$(\zeta_{\mathbf{Q}}(s))^2 \zeta_K(s) = \zeta_{k_1}(s) \zeta_{k_2}(s) \zeta_{k_3}(s). \tag{45}$$

We could check in the same way that the factors from the ramified primes also cancel, but it is less effort to argue as follows. Let $\theta(s)$ be the quotient of the two sides of (45), which we know to be the product of finitely many factors of the form $(1 - p^{-fs})^{\pm 1}$. Let $\psi(s)$ be the quotient of the products of expressions (77) below corresponding to the two sides of (45); then

$$1 = \frac{\psi(1 - s)}{\psi(s)} = \frac{\theta(1 - s)}{\theta(s)} |d/d_1 d_2 d_3|^{(1-2s)/2}$$

where the left hand equality follows from the functional equation and the right hand equality holds because the terms in curly brackets in (77) cancel in ψ. In view of the shape of $\theta(s)$, this can only happen if everything cancels; so $\theta(s) = 1$ and up to sign

$$d = d_1 d_2 d_3. \tag{46}$$

It is easy to check that the sign is also correct. Applying (75) to (45) gives

$$hR = h_1 h_2 h_3 R_1 R_2 R_3 \times (4w/w_1 w_2 w_3). \tag{47}$$

The last factor is easy to calculate, for $\sqrt[n]{1}$ can only lie in K if $\phi(n) = 1, 2$ or 4 where ϕ is Euler's function, by the Corollary to Theorem 27 below. The only even values of n satisfying this are those with $n \leqslant 12$; and we can rule out $n = 10$ because $\mathrm{Gal}(\mathbf{Q}(\sqrt[10]{1})/\mathbf{Q}) \approx C_4$. Now

$$n = 8 \iff K = \mathbf{Q}(\sqrt{-1}, \sqrt{2}) \quad \text{whence} \quad 4w/w_1 w_2 w_3 = 2;$$

in all other cases, including $n = 12$ for which $K = \mathbf{Q}(\sqrt{-1}, \sqrt{-3})$, it is easy to check that $4w/w_1 w_2 w_3 = 1$.

Now let η be a unit in K and denote by σ_i the non-trivial element of $\mathrm{Gal}(K/k_i)$. Then $\eta \cdot \sigma_i \eta$ is in k_i and therefore in $\mathfrak{o}_i^*$, the group of units of k_i; and on multiplying these three expressions together we see that

$$\eta^2 = \pm \eta^2 \mathrm{norm}_{K/\mathbf{Q}} \eta = \pm \prod (\eta \cdot \sigma_i \eta)$$

is in $\mathfrak{o}_1^* \mathfrak{o}_2^* \mathfrak{o}_3^*$. Denote by W the group of roots of unity in K; then it follows at once that

$$\epsilon = [\mathfrak{O}_K^* : W\mathfrak{o}_1^*\mathfrak{o}_2^*\mathfrak{o}_3^*] = \begin{cases} 1 \text{ or } 2 \text{ if } K \text{ is complex,} \\ 1, 2, 4 \text{ or } 8 \text{ if } K \text{ is real.} \end{cases} \tag{48}$$

Here we can delete W except when $K = \mathbf{Q}(\sqrt{-1}, \sqrt{2})$.

Suppose first that K is complex; choose the notation so that k_1 is real and let η be a fundamental unit of K. Then $\omega\eta^\epsilon$ is a fundamental unit for k_1 where ω is a root of unity in K, and hence $R_1 = \frac{1}{2}\epsilon R$; taking into account the anomalous case $K = \mathbf{Q}(\sqrt{-1}, \sqrt{2})$, (47) gives

$$h = \tfrac{1}{2}h_1h_2h_3[\mathfrak{O}_K^* : \mathfrak{o}_1^*\mathfrak{o}_2^*\mathfrak{o}_3^*].$$

If instead K is real, a similar calculation gives

$$h = \tfrac{1}{4}h_1h_2h_3[\mathfrak{O}_K^* : \mathfrak{o}_1^*\mathfrak{o}_2^*\mathfrak{o}_3^*].$$

I know of no other way to prove such results; and study of particular cases suggests that there is no corresponding relation between the ideal class groups.

13 Cyclotomic fields

Let k_m be the field of m-th roots of unity. By a **cyclotomic field** I shall mean any subfield of any k_m; but the reader is warned that some authors restrict the phrase to the k_m themselves. It is in any case natural to start with the k_m themselves; the properties of their subfields can then be deduced by means of Theorem 16. The most interesting fact about a general cyclotomic field is that one can write down explicitly units of a particular kind (the so-called cyclotomic units) and that usually these generate a subgroup of finite index in its full group of units. Moreover, this index is the product of the ideal class number h and factors which are easy to compute.

In studying the k_m it is convenient to deal with the prime factors of m one at a time — that is, to deal first with the special case when m is a prime power and then, writing $m = \prod p_i^{r_i}$, to obtain the field k_m by composition from the various fields k_{p^r} using Lemma 25 and its Corollary.

Theorem 27 *Let ρ be a primitive p^r-th root of unity, where p is prime; then $N = [\mathbf{Q}(\rho) : \mathbf{Q}] = p^{r-1}(p-1)$ and the conjugates of ρ over $\mathbf{Q}$ are the distinct ρ^n with n prime to p. The numbers $1, \rho, \ldots, \rho^{N-1}$ form a base for the integers of $\mathbf{Q}(\rho)$. Write $M = p^{r-1}(pr - r - 1)$; then the different of*

$\mathbf{Q}(\rho)$ is $(1 - \rho)^M$ and its discriminant is $\pm p^M$ where the sign is negative if $p \equiv 3 \bmod 4$ or $p^r = 4$ and positive otherwise. The roots of unity in $\mathbf{Q}(\rho)$ are just the $\pm\rho^\mu$. The only prime which ramifies is $(p) = ((1 - \rho))^N$. There is a natural isomorphism between $\mathrm{Gal}(\mathbf{Q}(\rho)/\mathbf{Q})$ and the multiplicative group of residue classes $\bmod\, p^r$ prime to p, where the automorphism corresponding to $n \bmod p^r$ is $\sigma_n : \rho \mapsto \rho^n$. If ℓ is a prime other than p, the Artin symbol is $\left(\frac{\mathbf{Q}(\rho)/\mathbf{Q}}{(\ell)}\right) = \sigma_\ell$.

Proof Certainly ρ satisfies the equation

$$\psi(X) = \frac{X^{p^r} - 1}{X^{p^{r-1}} - 1} = 0$$

of degree N, whose roots are precisely the ρ^n with n prime to p. In particular

$$\prod (1 - \rho^n) = \psi(1) = p. \tag{49}$$

But $1 - \rho$ divides $1 - \rho^n$; and if n' is such that $nn' \equiv 1 \bmod p^r$ then $1 - \rho^n$ divides $1 - \rho^{nn'} = 1 - \rho$. Hence each factor in the product (49) is $1 - \rho$ times a unit, and in terms of ideals we have $(p) = ((1 - \rho))^N$. This implies that $[\mathbf{Q}(\rho) : \mathbf{Q}] \geqslant N$; and since consideration of ψ gives the opposite inequality it follows that $[\mathbf{Q}(\rho) : \mathbf{Q}] = N$, that ψ is irreducible over $\mathbf{Q}$, that all the ρ^n with n prime to p are conjugate to ρ, and that $1 - \rho$ is a prime ideal. Since an element of the Galois group is determined by its action on ρ, the Galois group consists of the $\sigma_n : \rho \mapsto \rho^n$ with n prime to p and $\sigma_{mn}\rho = \rho^{mn} = \sigma_m\sigma_n\rho$.

Except in the special case $p^r = 2$ all embeddings of $\mathbf{Q}(\rho)$ in $\mathbf{C}$ are complex, and therefore $r_1 = 0$ and $r_2 = \frac{1}{2}N$; moreover we know that the sign of the discriminant is $(-1)^{r_2}$. Since

$$\prod (\rho - \sigma\rho) = \psi'(\rho) = p^r \rho^{p^{r-1}-1}/(\rho^{p^{r-1}} - 1)$$

where the product is taken over all the $\sigma\rho$ other than ρ itself, and

$$\mathrm{norm}_{\mathbf{Q}(\rho)/\mathbf{Q}}(\rho^{p^{r-1}} - 1) = \pm(\mathrm{norm}_{\mathbf{Q}(\sqrt[p]{1})/\mathbf{Q}}(\sqrt[p]{1} - 1))^{p^{r-1}} = \pm p^{p^{r-1}}$$

by (49) in the special case $r = 1$, we find that

$$\Delta^2(1, \rho, \dots, \rho^{N-1}) = \pm\mathrm{norm}\,\psi'(\rho) = \pm p^M,$$

which is the value asserted for the discriminant up to sign. In particular, the only factor we can take out of this is a power of p. To assert that the ρ^n for $0 \leqslant n < N$ form a base for the integers is the same as to assert that the $(1 - \rho)^n$ form one. Thus to prove the assertions about the discriminant and a base for the integers, it is enough to show that if the a_n are in $\mathbf{Q}$ for $0 \leqslant n < N$ and are such that $\sum a_n (1 - \rho)^n$ is an integer then the a_n are integers. But all the non-zero terms in this last sum have distinct $(1 - \rho)$-adic additive valuations, so each of them must be an integer. Since there is only one prime ideal in $\mathbf{Q}(\rho)$ which divides (p), the different is uniquely determined by the fact that its Norm is the ideal generated by the discriminant. Now suppose that there is a root of unity ϵ in $\mathbf{Q}(\rho)$ which is not of the form $\pm \rho^\mu$; raising it to a power if necessary we can assume it is an ℓ^s-th root of unity for some prime ℓ. Here $d_{\mathbf{Q}(\epsilon)} | d_{\mathbf{Q}(\rho)}$ because $\mathbf{Q}(\epsilon) \subset \mathbf{Q}(\rho)$; and since we have already shown that $d_{\mathbf{Q}(\epsilon)}$ is divisible by ℓ this implies $\ell = p$. Comparing discriminants now gives $s \leqslant r$, contrary to hypothesis.

Finally, the Artin symbol satisfies $(\frac{\mathbf{Q}(\rho)/\mathbf{Q}}{(\ell)})\xi \equiv \xi^\ell \bmod \ell$ for every integer ξ in $\mathbf{Q}(\rho)$. Taking $\xi = \rho$, it is evident that σ_ℓ is the only element of the Galois group which meets this condition. $\qquad\square$

Corollary *Let ρ be a primitive m-th root of unity for some $m > 2$; then $[\mathbf{Q}(\rho) : \mathbf{Q}] = \phi(m)$ where ϕ is Euler's function. If n is prime to m then ρ^n is conjugate to ρ over $\mathbf{Q}$. The roots of unity in $\mathbf{Q}(\rho)$ are just the $\pm \rho^\mu$. The primes which ramify in $\mathbf{Q}(\rho)$ are just those which divide m, except that 2 does not ramify if $2\|m$. $\mathrm{Gal}(\mathbf{Q}(\rho)/\mathbf{Q})$ is isomorphic to the multiplicative group of residue classes $\bmod\, m$ prime to m, where the element corresponding to $n \bmod m$ is $\sigma_n : \rho \mapsto \rho^n$. If ℓ is a prime not dividing m the Artin symbol is given by $(\frac{\mathbf{Q}(\rho)/\mathbf{Q}}{(\ell)}) = \sigma_\ell$.*

Proof Recall that if $m = \prod p_i^{r_i}$ then the value of Euler's function is

$$\phi(m) = m \prod_{p|m} \left(\frac{p-1}{p} \right).$$

Now everything follows at once from the theorem, together with Lemma 25 and its Corollary. These also give us the value of the different and hence of the discriminant. The assertion about the roots of unity is proved in the same way as in the theorem. $\qquad\square$

For any prime p, let $m = p^r m'$ where m' is prime to p. If A_m denotes the group of residue classes $\bmod\, m$ prime to m, there is a canonical isomorphism

$$\mathrm{Gal}(k_m/\mathbf{Q}) \approx A_m \approx A_{m'} \times A_{p^r} \approx \mathrm{Gal}(k_{m'}/\mathbf{Q}) \times \mathrm{Gal}(k_{p^r}/\mathbf{Q})$$

and the primes above p in $k_{m'}$ are totally ramified in $k_m/k_{m'}$. It follows that the splitting group of p is $C \times A_{p^r}$, where $C \subset A_{m'}$ is the cyclic group generated by p, and the inertia group of p is A_{p^r}. The corresponding statements for any cyclotomic field now follow from Theorem 16.

The following result will be needed both in §13.2 and in §20.

Lemma 30 *Let $k = \mathbf{Q}(\zeta)$ where ζ is a primitive p-th root of unity, and write $\pi = 1 - \zeta$. If ξ is in $\mathfrak{o}_k$ and prime to π, and if $\xi \equiv \alpha_0^p \bmod \pi^p$ for some α_0 in $\mathfrak{o}_{(\pi)}^*$, then (π) is unramified in K/k where $K = k(\sqrt[p]{\xi})$.*

Proof Suppose first that $\xi \equiv \alpha_0^p \bmod \pi^{p+1}$. In the notation of Lemma 20 we have $m = p - 1$, so that we can take $r = 2$; hence $\xi = \alpha^p$ for some α in $\mathfrak{o}_{(\pi)}^*$, so that (π) splits completely in K by Theorem 19. If instead $\pi^p \| (\xi - \alpha_0^p)$, we can assume that $K \neq k$, which indeed follows from the last assumption. Let $\eta = (\sqrt[p]{\xi} - \alpha_0)/\pi$, so that $K = k(\eta)$, and let $f(X)$ be the minimal monic polynomial for η; then

$$f(X) \equiv X^p + (\alpha_0^{p-1} p/\pi^{p-1})X + (\alpha_0^p - \xi)/\pi^p \bmod \pi.$$

It follows that η is in $\mathfrak{O}_K$; hence $\mathfrak{d}_{K/k}$ is prime to π by Theorem 20, and so (π) is unramified in K/k by Theorem 21. It is not hard to show that in this case (π) remains prime in K.   □

13.1 Class numbers of cyclotomic fields

Let k be any cyclotomic field with $[k : \mathbf{Q}] = n$ and let m be such that $k \subset k_m$; it is proved in §20 that all fields abelian over $\mathbf{Q}$ can be obtained in this way. Let A_m be the group of residue classes $\bmod\, m$ prime to m and H_m the kernel of the map $A_m \to \mathrm{Gal}(k_m/\mathbf{Q}) \to \mathrm{Gal}(k/\mathbf{Q})$ given by the Corollary to Theorem 27; we shall meet almost the same notation again in §17. Let χ be any character of A_m/H_m and let $\mathfrak{f}_\chi$ be the exact conductor of χ — that is, the least $\mathfrak{f}$ such that χ is defined $\bmod \mathfrak{f}$; and write $f_\chi = \mathrm{Norm}\,\mathfrak{f}_\chi$. It will follow from Theorem 33 that $|d_k| = \prod f_\chi$; alternatively, we can apply to both sides of (50) the functional equations implicitly described in (79) and (77) below. The key to what follows is the identity

$$\zeta_k(s) = \prod_\chi L(s, \chi), \tag{50}$$

which follows from the remarks after the proof of the Corollary to Theorem 27. If χ_0 is the principal character, $L(s, \chi_0)$ is just the Riemann zeta function (71). Multiplying both sides of (50) by $s - 1$, letting $s \to 1$ and using (75) we obtain

$$hR = \begin{cases} 2^{1-n}|d_k|^{1/2}\prod'L(1,\chi) & \text{if } k \text{ is real,} \\ (2\pi)^{-n/2}w|d_k|^{1/2}\prod'L(1,\chi) & \text{if } k \text{ is not real,} \end{cases} \tag{51}$$

where in each case the product is over all characters of A_m/H_m except χ_0. To justify this, we must show that the $L(1,\chi)$ are well-defined; and to make it useful we must obtain a closed formula for the $L(1,\chi)$, which we now do. Let x run through a set of representatives of the residue classes mod f_χ and let y run through those prime to f_χ. If ζ is a primitive f_χ-th root of unity then

$$L(s,\chi) = \sum_y \chi(y) \sum_{n=1}^{\infty} n^{-s} f_\chi^{-1} \sum_x \zeta^{(y-n)x}$$

because the innermost sum is f_χ if $n \equiv y \bmod f_\chi$ and 0 otherwise. By rearranging we obtain

$$L(s,\chi) = f_\chi^{-1}\sum_x \left\{ \sum_y \chi(y)\zeta^{xy} \right\} \left\{ \sum_{n=1}^{\infty} n^{-s}\zeta^{-nx} \right\}. \tag{52}$$

If $x \equiv 0 \bmod f_\chi$ the first expression in curly brackets vanishes. But if $x \not\equiv 0 \bmod f_\chi$ then at least formally

$$\lim_{s \to 1} \sum n^{-s}\zeta^{-nx} = \sum (\zeta^{-nx}/n) = -\log(1 - \zeta^{-x}) \tag{53}$$

where we choose that branch of the logarithm which has imaginary part strictly between $-\pi i$ and πi. This last step can be justified by a standard theorem of Abel or by the exercise which follows.

Exercise If $|z| < 1$, show by uniform convergence that

$$\lim_{s \to 1} \sum n^{-s}z^n = \sum n^{-1}z^n = -\log(1 - z).$$

If also $|z_1| \leqslant 1$ show that

$$\phi(s,z_1) = \sum n^{-s}z_1^n = \sum_{m=1}^{\infty} \{m^{-s} - (m+1)^{-s}\}(z_1 + \cdots + z_1^m)$$

and that, provided $|z_1 - 1| \geqslant \epsilon$ for some fixed $\epsilon > 0$, the right hand side is absolutely convergent uniformly in z_1 and s provided $\Re s \geqslant \epsilon$. Deduce that $\lim_{s \to 1} \phi(s, z_1)$ exists and is continuous in z_1, and derive (53). $\qquad \square$

The Gauss sum, which is the first expression in curly brackets in (52), satisfies

$$\tau_x(\chi) = \sum \chi(y)\zeta^{xy} = \overline{\chi}(x)\tau_1(\chi). \tag{54}$$

For if x is not prime to f let r be the quotient of f by the highest common factor of x and f. Replace y by yc where $c \equiv 1 \bmod r$; since $cx \equiv x \bmod f$ we obtain $\tau_x(\chi) = \chi(c)\tau_x(\chi)$. Thus $\tau_x(\chi) = 0$ because if $\chi(c) = 1$ for all such c, the conductor of χ would divide r. If on the other hand x is prime to f then $\tau_x(\chi)\chi(x) = \sum_y \chi(xy)\zeta^{xy} = \tau_1(\chi)$. Moreover

$$\tau_1(\chi)\overline{\tau_1(\chi)} = \sum_x \sum_y \chi(x)\overline{\chi}(y)\zeta^{x-y} = \sum_y \sum_z \chi(z)\zeta^{y(z-1)}$$

on writing $x = yz$. Summing first over y gives $|\tau_1(\chi)|^2 = f$.

It follows from (53) and (54) that

$$L(1,\chi) = -f_\chi^{-1} \sum \tau_x(\chi)\log(1 - \zeta^{-x}) = -f_\chi^{-1}\tau_1(\chi)\sum \overline{\chi}(x)\log(1 - \zeta^{-x}).$$

Since $\prod f_\chi = |d_k|$, the contribution of the $f_\chi^{-1}\tau_1(\chi)$ to $\prod'|L(1,\chi)|$ will be $|d_k|^{-1/2}$ which will cancel with the $|d_k|^{1/2}$ in (51).

It is now necessary to split cases. We shall say that χ is an **even** character if $\chi(-1) = 1$ and an **odd** character if $\chi(-1) = -1$. Since complex conjugacy on k is induced by $\rho \mapsto \rho^{-1}$ where ρ is a primitive m-th root of unity, it follows from the Corollary to Theorem 27 that all characters are even if k is real but half of them are odd if k is not real. If k is not real, it has a totally real subfield k_0 such that $[k : k_0] = 2$; this is the field fixed by complex conjugacy, so the characters associated with k_0 are just the even characters associated with k. It follows from (51) that if k is real

$$2^{n-1}hR = \prod' \left| \sum \chi(x)\log(1 - \zeta^{-x}) \right|, \tag{55}$$

whereas if k is not real

$$h = 2h_0\{2^{n/2-2}wR_0/R\}(2\pi)^{-n/2}\prod'' \left| \sum \chi(x)\log(1 - \zeta^{-x}) \right| \tag{56}$$

where h_0 and R_0 are the values for k_0, the product is taken over all odd characters and the expression in curly brackets is equal to $[\mathfrak{o}_k^* : \mathfrak{o}_{k_0}^*]$. For further information on this last expression, see Exercises 1.8 and 3.9.

Suppose first that k is real. Let f be a factor of m; for what follows to be non-trivial, f must be a multiple of f_χ for some non-principal χ. Let H_f be the image of H_m in A_f and ζ a primitive f-th root of unity. Fix some a prime to f and not in H_f, and let c run through a half-set of representatives of the classes $\bmod f$ in H_f, so that the $c \bmod f$ and the $-c \bmod f$ together represent all the classes in H_f. Then

$$\prod_c{}' \{\zeta^{c(1-a)/2}(1 - \zeta^{ac})/(1 - \zeta^c)\} \tag{57}$$

is a unit in $\mathbf{Q}(\zeta)$ and is fixed by $\mathrm{Gal}(\mathbf{Q}(\zeta)/k \cap \mathbf{Q}(\zeta)) \approx H_f$; so it is a unit in k, and it is not obviously trivial. The units of k composed from -1 and these units are called the **cyclotomic units** of k. In general, it follows from Theorem 11 that there are multiplicative relations among the units (57) as a and f vary; but no one has sorted out the details. One should expect that the right hand side of (55) is the regulator of a suitable set of cyclotomic units; but this again is unproved. Both these are problems which defeated Hasse, so they will not be easy. All that one knows is the following.

Let χ be a character $\bmod f$, but not necessarily with conductor f, let ζ be an f-th root of unity, and write

$$S(\chi, \zeta) = \sum \chi(x) \log(1 - \zeta^{-x}) \tag{58}$$

where the sum is taken over a set of representatives x of the residue classes $\bmod f$ prime to f. For any prime p we can obtain a character χ^* $\bmod fp$ by restricting the argument of χ to be prime to p. Let $\xi^p = \zeta$; then

$$(1 - \zeta^{-x}) = (1 - \xi^{-x})(1 - \xi^{-x-f}) \cdots (1 - \xi^{-x-(p-1)f}). \tag{59}$$

If $p|f$ then substituting (59) into (58) gives $S(\chi, \zeta) = S(\chi^*, \xi)$. If however $p \nmid f$ then we can write $x = py$ in (58) and obtain

$$S(\chi, \zeta) = \sum \chi(py) \log(1 - \xi^{-py}) + S(\chi^*, \xi) = \chi(p) S(\chi, \zeta) + S(\chi^*, \xi).$$

Thus in either case we have

$$S(\chi^*, \xi) = (1 - \chi(p)) S(\chi, \zeta).$$

We can now rewrite (55) as

$$hR \left\{ \prod \prod (1 - \chi(p)) \right\} = {\prod}' \left| \tfrac{1}{2} \sum \chi^*(x) \log(1 - \zeta_m^{-x}) \right|. \tag{60}$$

Here ζ_m is a primitive m-th root of unity, the double product on the left is taken over all primitive characters χ induced by characters χ^* of A_m/H_m and over all distinct primes $p|m$, and the sum on the right is taken over a complete set of representatives x for the elements of A_m/H_m. Of course, this formula is only useful if no $\chi(p)$ is equal to 1.

The right hand side of (60) can be expressed as a regulator by means of the second part of the following lemma, due to Dirichlet.

Lemma 31 *With the notation above, let the $B(y)$ be variables indexed by the elements of A/H. Then*

$$\prod \left\{ \sum_j \chi_i^*(y_j) B(y_j) \right\} = \det(B(y_i y_j^{-1})) \tag{61}$$

where the χ_i^* *run through the characters of* A/H *and* y_i, y_j *run through the elements of* A/H*; and*

$$\prod{}' \left\{ \sum \chi_i^*(y_j) B(y_j) \right\} = \det(B(y_i y_j^{-1}) - B(y_i)) \qquad (62)$$

where the product on the left excludes the principal character and y_i, y_j *on the right run through the elements of* A/H *other than* H *itself.*

Proof If we multiply the i-th row of the determinant in (61) by $\chi^*(y_i)$ and add all the other rows to the first one, the term in the j-th column becomes $\chi^*(y_j) \sum \chi^*(y) B(y)$; so the right hand side of (61), considered as a polynomial in the $B(y)$, is divisible by $\sum \chi^*(y) B(y)$. Since this holds for each χ^*, the right hand side of (61) is divisible by the left hand side. By considering degrees, the quotient must be a constant; and by looking at the coefficient of $(B(H))^n$ that constant must be 1.

We can renumber the y_i so that $y_1 = H$. By the argument in the previous paragraph, the left hand side of (62) is equal to the determinant obtained from that in (61) by replacing the first row by $(1, 1, \ldots , 1)$. Now subtract the first column from each of the others.   □

In each sum on the right of (60) we can write $x = cy$, where y runs through representatives of the elements of A/H and c runs through representatives of the congruence classes mod f in H. Thus

$$\sum{}_x \chi^*(x) \log(1 - \zeta_m^{-x}) = \sum{}_y \chi^*(y) \sum{}_c \log(1 - \zeta_m^{-cy}). \qquad (63)$$

Let σ_j^{-1} be the element of $\mathrm{Gal}(k/\mathbf{Q})$ corresponding to the class of y_j in A_m/H_m. Write

$$\eta_j = \prod{}'_c \left\{ (1 - \sigma_j \xi^c) \sqrt{\sigma_j \xi^c / \xi^c} / (1 - \xi^c) \right\} = \sqrt{\prod{}_c \{ (1 - \sigma_j \xi^c)/(1 - \xi^c) \}}$$

where $\prod'$ is taken over a half-set of representatives of the classes mod m in H_m and $\prod$ is taken over a full set of representatives. Since $\sqrt{\sigma_j \xi / \xi}$ is an integral power of ξ, the η_j are cyclotomic units in k, and $\eta_1 = 1$ since σ_1 is the identity. It now follows from (62) and (63) that the right hand side of (60) is the regulator of $\eta_2, \ldots , \eta_n$.

We now turn to (56). Here every χ is odd, so that

$$\sum \chi(x) \log(1 - \zeta^{-x}) = \tfrac{1}{2} \sum \chi(x) \{ \log(1 - \zeta^{-x}) - \log(1 - \zeta^x) \}$$
$$= -\pi i f^{-1} \sum{}' x \chi(x)$$

where x runs through all the integers prime to f with $|x| < \tfrac{1}{2} f$. Taken with (56), this gives a formula for h/h_0 in which all the terms are rational and easy to compute. That h/h_0 is actually integral is an easy consequence of class field theory; see Exercise 5.3. There are elementary proofs of this last statement, which are straightforward when $k = k_m$ but complicated in general.

13.2 Fermat's Last Theorem

A major motivation for studying k_n was the possibility of applications to Fermat's Last Theorem — the assertion that

$$X^n + Y^n = Z^n, \quad XYZ \neq 0$$

has no solution in rational integers when $n > 2$. Since any integer greater than 2 is divisible either by 4 or by an odd prime, it is enough to consider the cases when n is either 4 or an odd prime p. The cases $n = 3$ and $n = 4$ were proved by Fermat; for two versions of the argument when $n = 4$, see Exercises 3.5 and 3.6. Although there is no way of knowing, it seems likely that the argument which convinced Fermat in the general case was of the same type as that given below, combined with the assumption of unique factorization. The oldest recorded version of this argument is due to Kummer in the mid-nineteenth century. He was able to make it rigorous, but the price was the additional condition that p should be a so-called 'regular' prime — that is to say, that the class number of $\mathbf{Q}(\sqrt[p]{1})$ is prime to p. Though most small primes (including all $p < 37$) are regular it is still not even known whether there are infinitely many regular primes.

It is enough to assume that there are integers x, y, z with highest common factor 1 such that

$$x^p + y^p = z^p, \quad xyz \neq 0 \tag{64}$$

and derive a contradiction. Denote by ζ a primitive p-th root of unity and write $\pi = 1 - \zeta$. We separate cases according as $p \nmid xyz$ or $p \mid xyz$.

In the first case we can take $p \geqslant 5$, for if $p = 3$ each of x^p, y^p, z^p would be congruent to $\pm 1 \bmod 9$ and this contradicts the first equation (64). If $p \geqslant 5$ we write (64) in the form

$$(x + y)(x + \zeta y) \cdots (x + \zeta^{p-1} y) = z^p; \tag{65}$$

the factors on the left are coprime in pairs because they can have no common factor other than (π), and by hypothesis (π) does not divide the right hand side. Hence each of these factors, viewed as an ideal, must be a p-th power. But since h is prime to p, an ideal whose p-th power is principal must itself be principal. Hence each factor on the left in (65), regarded as a number, must be the product of a p-th power and a unit. In particular

$$x + \zeta y = \alpha^p u \tag{66}$$

for some integer α prime to π and some unit u. If we use a bar to denote complex conjugacy, then $u/\bar{u}$ is a unit all of whose conjugates over $\mathbf{Q}$ have absolute value 1; thus it is a root of unity and must have the form $\eta = \pm\zeta^m$ for some m with $0 \leqslant m < p$. Also $\alpha \equiv \bar{\alpha} \bmod \pi$ by considering the expression for α as an element of $\mathbf{Z}[\rho]$, and therefore $(\alpha/\bar{\alpha})^p \equiv 1 \bmod p\pi$. Comparing (66) with its complex conjugate, we obtain

$$x + \zeta y - \eta(x + \zeta^{-1}y) \equiv 0 \bmod p\pi.$$

Multiplying this congruence by ζ if $m = 0$, or by ζ^2 if $m = p - 1$, we see that if $m \neq 1$ there is a polynomial $f(T)$ in $\mathbf{Z}[T]$ of degree at most $p-2$, not divisible by p but such that $f(\zeta) \equiv 0 \bmod p\pi$. But now $g(U) = f(1-U)$ has degree at most $p - 2$ and is not divisible by p, but $g(\pi) \equiv 0 \bmod p\pi$; and this is impossible because the terms in $g(\pi)$ all have distinct valuations. Thus $m = 1$ and $\eta = \pm\zeta$. But now $y \equiv \pm x \bmod p$. If the lower sign held, then $x + y$ would be divisible by p, whence $p|z$ contrary to hypothesis; so $x \equiv y \bmod p$.

Applying the same argument to $(-x)^p + z^p = y^p$ gives $-x \equiv z \bmod p$. Now substituting back into (64) gives $p = 3$, and this has already been ruled out. This completes the discussion of the first case.

Nearly all the difficulty in the second case is in the proof of the following fundamental lemma. This was another of the precursors of class field theory. An elementary proof, which goes back to Kummer, is sketched in the exercise later in this subsection; in the text we derive the result painlessly from classical class field theory.

Lemma 32 *In the notation above, let ϵ be a unit of $k = \mathbf{Q}(\zeta)$ such that $\epsilon \equiv \alpha_1^p \bmod \pi^p$ for some α_1 in $\mathfrak{o}_k$. Then either $\epsilon = \xi^p$ for some unit ξ in k or $p|h$.*

Proof Suppose that the first alternative does not happen. The p roots of $X^p - \epsilon$ form complete sets of conjugates, all sets being of equal size; so they are all conjugate and $K = k(\sqrt[p]{\epsilon})$ is abelian of degree p over k. But $\mathfrak{d}_{K/k}$ divides p, by Theorem 20 applied to $\alpha = \sqrt[p]{\epsilon}$. Lemma 30 shows that (π) is unramified in K/k; so K/k is not ramified at any place. In the terminology of §17 this means that H contains the group of principal ideals, so the ideal class group has a quotient group isomorphic to $\mathrm{Gal}(K/k)$, whence $p|h$. $\qquad\square$

For the second case of Fermat's Last Theorem, again under the condition that $p \nmid h$, it is convenient to consider a more general equation. (We shall use the method of infinite descent, invented by Fermat himself, and for this it is essential to choose an equation which will descend to itself.) Suppose that there are non-zero integers x, y, z in k, with π dividing z but not x or y, which give a solution of

$$X^p + Y^p = \epsilon Z^p \tag{67}$$

for some unit ϵ in k. Define n by $\pi^n \| z$, and write (67) in the form

$$(x + y)(x + \zeta y) \cdots (x + \zeta^{p-1} y) = \epsilon z^p. \tag{68}$$

At least one of the factors on the left in (68) must be divisible by π; so all of them are, and the highest common factor of any two is $\pi\mathfrak{a}$ where $\mathfrak{a} = (x, y)$ is prime to (π). Hence the factors on the left use up the p residue classes mod π^2 divisible by π, so that one of them is divisible by π^2; multiplying y by a power of ζ, we can take this factor to be $x + y$. Thus $\pi \| (x + \zeta^r y)$ if $p \nmid r$, and $\pi^{p(n-1)+1} \| (x + y)$; in particular $n > 1$.

The product of the ideals associated with the factors on the left of (68) is a p-th power, and the highest common factor of any two of them is $\pi\mathfrak{a}$; so as ideals

$$(x + \zeta^r y) = \pi\mathfrak{a}\mathfrak{b}_r^p \quad \text{for } r = 0, 1, \ldots, p - 1. \tag{69}$$

Let $\mathfrak{c}$ be an integral ideal in the class of $\mathfrak{b}_0^{-1}$, divisible by $\mathfrak{a}$ but not by (π). Thus $\mathfrak{a}^{-1}\mathfrak{c}^p$ is principal and hence equal to (β) for some β in $\mathfrak{o}_k$, so that (69) becomes

$$(\beta x + \zeta^r \beta y) = \pi(\mathfrak{c}\mathfrak{b}_r)^p.$$

Since h is prime to p and $(\mathfrak{c}\mathfrak{b}_r)^p$ is a principal ideal, so is $\mathfrak{c}\mathfrak{b}_r$; if it is equal to (α_r) then

$$\beta x + \zeta^r \beta y = \pi\epsilon_r \alpha_r^p$$

for some units ϵ_r. Here $\pi^{n-1} \| \alpha_0$ and the other α_r are prime to π. Combining the equations with $r = 0, 1$ and $p - 1$ gives

$$\epsilon_1 \alpha_1^p + \zeta \epsilon_{-1} \alpha_{-1}^p = (1 + \zeta)\epsilon_0 \alpha_0^p.$$

This implies that $\zeta\epsilon_{-1}\epsilon_1^{-1} \equiv (-\alpha_1/\alpha_{-1})^p \bmod \pi^p$, so $\zeta\epsilon_{-1}\epsilon_1^{-1} = \eta^p$ for some unit η by Lemma 32. Thus

$$\alpha_1^p + (\eta\alpha_{-1})^p = (1 + \zeta)\epsilon_0\epsilon_1^{-1}\alpha_0^p,$$

and here $(1 + \zeta)\epsilon_0\epsilon_1^{-1}$ is a unit. So we have generated another solution of

(67), and indeed one with $\pi^{n-1}\|\alpha_0$. If we assume that x, y, z were chosen to make n as small as possible, we obtain a contradiction. This completes the discussion of the second case.

By a refinement of this argument, it can be shown that the second case is impossible under the weaker hypothesis $p \nmid h_0$, where h_0 is the class number of $\mathbf{Q}(\zeta + \zeta^{-1})$, the maximal real subfield of $\mathbf{Q}(\zeta)$. (That this hypothesis is weaker than $p \nmid h$ is a special case of Exercise 5.3.) Vandiver conjectured that p never divides h_0, and this has been verified by computer for all $p < 4 \times 10^6$. But naive probabilistic arguments suggest that counterexamples to Vandiver's conjecture should be so rare that no feasible amount of computation is likely to produce one.

Exercise The object of this exercise is to provide an elementary proof of Lemma 32. The first half consists of the following key result. It can be seen as a statement that a certain cohomology group is non-trivial — though this interpretation does not make the proof easier. We continue to denote the field of p-th roots of unity by k.

Lemma 33 *Let K be cyclic of degree p over k and let σ be a generator of $\mathrm{Gal}(K/k)$; then there is a unit η of K such that $\mathrm{norm}_{K/k}\eta = 1$ but η does not have the form $\epsilon/\sigma\epsilon$ for any unit ϵ of K.*

Note that if we did not require ϵ to be a unit, Lemma 4 would allow us to satisfy $\eta = \epsilon/\sigma\epsilon$. Write

$$U_1 = \mathfrak{O}_K^*/\mathfrak{o}_k^*, \quad U = U_1/\{\text{torsion part of } U_1\}.$$

It follows from Theorem 11 that

- U is a free abelian group on $\frac{1}{2}(p-1)^2$ generators.

Our terminology will reflect the fact that U is multiplicative. In particular, if $f(X) = \sum a_\nu X^\nu$ is in $\mathbf{Z}[X]$ we write $\alpha^{f(\sigma)} = \prod(\sigma^\nu\alpha)^{a_\nu}$ for any α in K. Let $F(X) = 1 + \cdots + X^{p-1}$ and recall that $F(X)$ is irreducible in $\mathbf{Z}[X]$ by Theorem 27; then $\theta^{F(\sigma)} = 1$ for every θ in U. Denote by $\mathcal{S}(\theta_1,\ldots,\theta_r)$ for any $\theta_1,\ldots,\theta_r$ in U the set of $\sigma^\nu\theta_\rho$ for $\nu = 0,\ldots,p-2$ and $\rho = 1,\ldots,r$. Write $m = \frac{1}{2}(p-1)$.

- We can find $\theta_1,\ldots,\theta_m$ in U such that the elements of $\mathcal{S}(\theta_1,\ldots,\theta_m)$ are multiplicatively independent.

Suppose that we have chosen $\theta_1,\ldots,\theta_r$ where $0 \leqslant r < m$, and let θ_{r+1} in

U be multiplicatively independent of $\mathcal{S}(\theta_1, \ldots, \theta_r)$. Suppose there were a relation of the form

$$\prod_{\rho=1}^{r+1} \theta_\rho^{f_\rho(\sigma)} = 1 \tag{70}$$

where the $f_\rho(X)$ are polynomials of degree at most $p - 2$ in $\mathbf{Z}[X]$ and $f_{r+1} \neq 0$. Then we could find $g(X), h(X)$ in $\mathbf{Z}[X]$ and $N \neq 0$ in $\mathbf{Z}$ with $gf_{r+1} + hF = N$. Raising (70) to the $g(\sigma)$-th power gives

$$\theta_{r+1}^N \prod_1^r \theta_\rho^{f_\rho(\sigma)g(\sigma)} = 1,$$

which contradicts the hypothesis on θ_{r+1}. It follows from the last two results that $\mathcal{S}(\theta_1, \ldots, \theta_m)$ spans a subgroup of finite index in U.

• If $\theta_1, \ldots, \theta_m$ are chosen so that the index of $\mathcal{S}(\theta_1, \ldots, \theta_m)$ in U is as small as possible then no θ_μ has the form $\theta^{1-\sigma}$ with θ in U..

Suppose otherwise. It is enough to show that θ is not in the subgroup generated by $\mathcal{S}(\theta_1, \ldots, \theta_m)$, because then replacing θ_μ by θ will replace this subgroup by a strictly larger one. Let

$$f(X) = (p-1) + (p-2)X + \cdots + X^{p-2}$$

so that $(1 - X)f(X) = p - F(X)$ and so $\theta^p = \theta_\mu^{f(\sigma)}$. If θ were in the subgroup generated by $\mathcal{S}(\theta_1, \ldots, \theta_m)$ we would have a contradiction. Now note that $\theta_\mu^p = (\theta_\mu^{f(\sigma)})^{1-\sigma}$ and deduce that if a is prime to p then $\theta_\mu^a \neq \theta^{1-\sigma}$. Applying an integral unimodular transformation, deduce:

• if $a_1, \ldots, a_m$ are integers not all divisible by p then $\prod \theta_\mu^{a_\mu}$ does not have the form $\theta^{1-\sigma}$ with θ in U.

Now let ζ be a primitive p-th root of unity. In proving the lemma we can assume $\zeta = \xi^{1-\sigma}$ for some unit ξ in K, for otherwise we can take $\eta = \zeta$. Hence $\xi^p = \operatorname{norm}_{K/k}\xi$ is a unit in k, and ξ is not in k. Now lift each θ_μ to a unit η_μ in K and consider the map

$$\phi : \{\text{group generated by the } \eta_\mu\} \to \mathfrak{o}_k^* / \{p\text{-th roots of unity}\}$$

induced by $\operatorname{norm}_{K/k}$. There are three possibilities:

(i) ξ^p is a primitive p-th root of unity. Let $\eta^\sharp$ be a primitive element in the kernel of ϕ, and choose η to be a product of $(\eta^\sharp)^a$ with a prime to p, a power of ξ and ± 1 such that $\operatorname{norm}_{K/k}\eta = 1$.

(ii) No non-trivial power of ξ^p is in the image of ϕ. The kernel of ϕ has rank at least 2, so that there is a primitive element $\eta^\sharp$ in the kernel of ϕ such that $\operatorname{norm}_{K/k}\eta^\sharp = \pm 1$. Take $\eta = \pm \eta^\sharp$.

(iii) Neither (i) nor (ii) holds. Now there is an element η^b in the group generated by the η_μ such that $(\mathrm{norm}_{K/k}\eta^b)/\xi^{pb}$ is a root of unity for some $b > 0$. Show that we can assume that η^b is not a p-th power as in (ii), using η^b/ξ^b and the kernel of ϕ.

This completes the proof of Lemma 33. For the proof of Lemma 32 let η be as in Lemma 33. We can assume that ϵ is not a p-th power in k; if we write $K = k(\sqrt[p]{\epsilon})$ then K/k is cyclic of degree p. It follows from Lemma 4 that there is an integer β in K such that $\eta = \beta^{1-\sigma}$. As an ideal, (β) is invariant under $\mathrm{Gal}(K/k)$. By considering the prime factorization of (β) and using the fact that there is no ramification in K/k by Lemma 30, show that $(\beta) = \mathrm{conorm}_{K/k}\mathfrak{b}$ for some ideal $\mathfrak{b}$ in k. For $\mathfrak{b}$ to be principal would contradict Lemma 33; on the other hand $\mathfrak{b}^p = \mathrm{norm}_{K/k}(\beta)$ is principal. $\square$

4

Analytic methods

14 Zeta functions and L-series

A fundamental tool in the study of prime numbers is the **Riemann zeta function**

$$\zeta(s) = \sum n^{-s} = \prod (1 - p^{-s})^{-1}, \qquad (71)$$

where the sum is over all integers $n > 0$ and the product is over all primes p; that the two are equal is equivalent to unique factorization. The sum and product are both absolutely convergent in $\Re s > 1$, and $\zeta(s) \sim (s-1)^{-1}$ as $s \to 1$ from the right. The first non-trivial property of $\zeta(s)$ is that it can be analytically continued to the entire s-plane, subject to a simple pole at $s = 1$, and that it satisfies the functional equation

$$\zeta(1 - s) = 2^{1-s}\pi^{-s}\Gamma(s)\zeta(s) \cos \tfrac{1}{2}\pi s;$$

this equation is equivalent to saying that

$$Z(s) = \Gamma(\tfrac{1}{2}s)\pi^{-s/2}\zeta(s) \qquad (72)$$

is unchanged by writing $1 - s$ for s. As we shall see below, the extra factor in (72) can be regarded as the missing factor in the product (71) corresponding to the infinite prime. The other fundamental property of $\zeta(s)$, assuming it is true, is the Riemann hypothesis: that the only zeros of $Z(s)$ lie on the line $\Re s = \tfrac{1}{2}$.

The problem of the distribution of primes in arithmetic progressions led to the study of the **Dirichlet L-series**

$$L(s, \chi) = \sum \chi(n) n^{-s} = \prod (1 - \chi(p)p^{-s})^{-1} \qquad (73)$$

where $\chi(n)$ is a **congruence character** mod f for some integer f — that is to say, $\chi(n) = 0$ when n is not prime to f, and if n is prime to f then χ is

79

induced by a character on the group of residue classes prime to f. $L(s, \chi)$ can also be analytically continued over the whole s-plane, though without a pole at $s = 1$ if χ is non-trivial, and it satisfies a functional equation of the form

$$L(1 - s, \chi) = f(s, \chi) L(s, \overline{\chi})$$

where $f(s, \chi)$ is a known and comparatively simple function. This equation cannot easily be put into a symmetric form; in fact $f(s, \chi)$ involves a Gauss sum and any symmetry would induce an identity between Gauss sums. There is also a Riemann hypothesis for $L(s, \chi)$.

Quite generally, to call a function a (global) zeta function or L-series is to assert that it is a Dirichlet series $\sum a_n n^{-s}$, and that after normalizing if necessary by writing $s - s_0$ for s it has four key properties:

(i) It can be written as an Euler product $\prod \phi_p(p^{-s})$ where each $\phi_p(X)$ is a rational function of X; note that this makes no allowance for factors corresponding to the infinite prime.

(ii) The series and product converge absolutely in $\Re s > 1$ but no further; and they can be analytically continued to the entire s-plane subject to possible poles at $s = 0$ and/or 1.

(iii) There is a functional equation connecting $L(s)$ with $\overline{L(1 - \bar{s})}$, where the latter expression is often obviously the same as $L(1 - s)$.

(iv) The non-trivial zeros of $L(s)$ lie on $\Re s = \frac{1}{2}$.

The function may also have zeros at $z = -n$ for certain $n \geqslant 0$; these should be thought of as the consequence of failing to include a factor corresponding to the infinite prime in the product required by (i). Fortunately, these four properties only need to be asserted as conjectures and not as proven facts; indeed modular forms give rise to examples where even (i) is not trivial, and there is as yet no case where (iv) has been proved. On the other hand, I know of no case in which the second part of (ii) has been proved without the argument simultaneously proving (iii). A function having these properties is usually called a zeta function if it has a pole at $s = 1$ and an L-series otherwise; but this is not a firm rule.

For an arbitrary algebraic number field k, the only possible analogue of the Riemann zeta function which satisfies (i) above is

$$\zeta_k(s) = \sum (\text{Norm } \mathfrak{a})^{-s} = \prod (1 - (\text{Norm } \mathfrak{p})^{-s})^{-1}, \qquad (74)$$

where the sum is over all non-zero integral ideals $\mathfrak{a}$ and the product is over all prime ideals $\mathfrak{p}$. Again, this identity is equivalent to unique factorization.

To any rational prime p there correspond at most $[k : \mathbf{Q}]$ primes $\mathfrak{p}$; so the product is absolutely convergent in $\Re s > 1$ and hence so is the sum.

It is reasonably easy to study the behaviour of $\zeta_k(s)$ as $s \to 1$ from the right. One rewrites the Dirichlet series as

$$\zeta_k(s) = \sum a_n n^{-s} = s^{-1} \int_0^\infty x^{-s-1} A(x) dx$$

where $A(x) = \sum_{n \leqslant x} a_n$; so it is enough to estimate $A(x)$. The calculation is essentially that in the proof of Theorem 23, and the conclusion is

$$(s - 1)\zeta_k(s) \to 2^{r_1}(2\pi)^{r_2}|d|^{-1/2} h R w^{-1} \quad \text{as } s \to 1. \tag{75}$$

We shall not carry out this process explicitly because the result is contained in Theorem 28 in the next section. The formula (75) is among other things the starting point of the proof of the Brauer-Siegel Theorem: that if k is Galois over $\mathbf{Q}$ then

$$\log(hR) \sim \tfrac{1}{2} \log |d| \quad \text{as } |d| \to \infty \tag{76}$$

for fields of any fixed degree; see [L], Chapter XVI. Empirical evidence suggests that if $r_1 + r_2 > 1$ (so that R is non-trivial) then $\log h$ is usually very small compared to $\log |d|$, though this is not so for the families of fields which one is most likely to write down; but it is very unlikely that there is anything analogous to (76) for h or R separately.

Analytic continuation and the functional equation for $\zeta_k(s)$ are much more difficult, and were only proved by Hecke in 1918 in a remarkable paper. With the advantages of hindsight, the basic idea of Hecke's paper is fairly natural; but the details are complicated and the proof depends on what appears at first sight to be a lucky accident. Hecke's first result was that

$$\zeta_k(s)\{\pi^{-s/2}\Gamma(\tfrac{1}{2}s)\}^{r_1}\{(2\pi)^{1-s}\Gamma(s)\}^{r_2}|d_k|^{s/2} \tag{77}$$

is unchanged on writing $1 - s$ for s. It is now natural to guess that $\pi^{-s/2}\Gamma(\tfrac{1}{2}s)$ and $(2\pi)^{1-s}\Gamma(s)$ are the missing factors in the product formula for $\zeta_k(s)$ corresponding to a real and a complex infinite place respectively, even though they look totally unlike the factors in (74) coming from the finite primes. We have already seen that the discriminant $d = \pm \text{Norm}\,\mathfrak{d}$ is conductor-like, so that the factor $|d_k|^{s/2} = (\text{Norm}\,\mathfrak{d})^{s/2}$ should be thought of as related to the conductor of the field k.

For L-series, the situation is more complicated. Choose a finite set S of prime ideals and denote by I^S the group of those fractional ideals whose prime factorizations do not involve any ideal in $\mathcal{S}$. Let χ be a character of $I^{\mathcal{S}}$ and write

$$L(s,\chi) = \sum \chi(\mathfrak{a})(\text{Norm } \mathfrak{a})^{-s} = \prod (1 - \chi(\mathfrak{p})(\text{Norm } \mathfrak{p})^{-s})^{-1} \qquad (78)$$

where the sum is restricted to integral ideals in $I^{\mathcal{S}}$ and the product to prime ideals not in $\mathcal{S}$. To prove analytic continuation and a functional equation, we need to impose certain conditions on χ which constitute the definition of a Hecke Grössencharakter. However, the appropriate factors for the infinite places are no longer those that appear in (72) and (77); we have to replace s by $s + s_v$, where s_v depends both on χ and on the particular place we are looking at. Replacing χ by χ^{-1} reverses the signs of all the s_v. Write

$$\Gamma(s,\chi) = \prod{}' \{\pi^{-(s+s_v)/2}\Gamma(\tfrac{1}{2}(s+s_v))\}\prod{}'' \{(2\pi)^{1-s-s_v}\Gamma(s+s_v)\}$$

where the first product is taken over the real and the second over the complex places. Hecke's second result was that

$$\Lambda(s,\chi) = L(s,\chi)\Gamma(s,\chi)\{\text{Norm}(\mathfrak{f}\mathfrak{d})\}^{s/2} \qquad (79)$$

is multiplied by a number of absolute value 1, depending only on χ, when s is replaced by $1 - s$ and χ by χ^{-1}. He expressed this last number as a finite sum, which generalizes the Gauss sum which appears in the classical case $k = \mathbf{Q}$.

Tate's thesis, which was written in 1950 but not published until 1967 (as Chapter XV of [CF]), presented a new proof of analytic continuation and the functional equation, using the language of adèles and idèles. General theory now replaces the heavy calculations needed by Hecke, but some at least of the ideas remain the same. Tate's proof is given in the next section.

A completely different generalization of the L-series (78) was introduced by Artin. To explain his motivation, we return to (73). Let K be the field of f-th roots of unity. By the Corollary to Theorem 27, we can interpret χ as a character on $\text{Gal}(K/\mathbf{Q})$; and (73) then becomes

$$L(s,\chi) = \prod \left(1 - \chi((\tfrac{K/\mathbf{Q}}{(p)}))p^{-s}\right)^{-1}$$

where $(\tfrac{K/\mathbf{Q}}{(p)})$ is the Artin element and the product is taken over all p prime to f. If we assume the assertions in §17, we can give an exactly analogous interpretation of (78) in terms of a certain abelian extension K/k. What Artin did was to consider extensions K/k which are Galois but not necessarily abelian. Instead of a character χ it is now necessary to use a **representation** ρ of $\text{Gal}(K/k)$ — that is, a homomorphism $\rho :$ $\text{Gal}(K/k) \to GL(W, \mathbf{C})$ where W is a finite dimensional vector space. If $\mathfrak{p}$ is a prime ideal in k unramified in K/k and $\mathfrak{P}$ is one of the primes of K above $\mathfrak{p}$, then the associated factor in the Artin L-series is

$$\{\det\{I - \rho([\tfrac{K/k}{\mathfrak{P}}])(\text{Norm}\,\mathfrak{p})^{-s}\}\}^{-1},$$

which does not depend on the choice of $\mathfrak{P}$ or of ρ within its equivalence class. We have still to take account of ramified primes. If $\mathfrak{p}$ is ramified in K/k then the Frobenius element $[\tfrac{K/k}{\mathfrak{p}}]$ is a coset of the inertia group $T_{\mathfrak{P}}$; so the space it naturally acts on is W^T, the subset of W whose elements are invariant under the action of T via ρ. Thus the L-series which we are led to consider is

$$L(s, \rho) = \prod_{\mathfrak{p}} \left\{ \det\{I - (\text{Norm}\,\mathfrak{p})^{-s}\rho([\tfrac{K/k}{\mathfrak{P}}]) \text{ acting on } W^T\} \right\}^{-1}.$$

This is the Artin L-series.

For any σ in $\text{Gal}(K/k)$ the characteristic roots of $\rho(\sigma)$ have absolute value 1; so the product is absolutely convergent in $\Re(s) > 1$. Clearly

$$L(s, \rho_1 + \rho_2) = L(s, \rho_1)L(s, \rho_2);$$

and if ρ is the principal representation (for which $\dim W = 1$ and every $\rho(\sigma)$ is the identity) then $L(s, \rho) = \zeta_k(s)$. It is known that $L(s, \rho)$ can be analytically continued to the whole s-plane as a meromorphic function and that it satisfies a functional equation of the usual type. It is conjectured that if ρ is irreducible and non-principal then $L(s, \rho)$ is everywhere holomorphic. The importance of this conjecture is that it would imply that $\zeta_K(s)/\zeta_k(s)$ is holomorphic for any algebraic number fields K, k with K Galois over k.

The expression for $\zeta_K(s)$ as a product tells us how rational primes p factorize in K. Suppose for convenience that K is normal over $\mathbf{Q}$ and let L be any field intermediate between $\mathbf{Q}$ and K. If we know how p factorizes in K then Theorem 16 tells us how it factorizes in L. Since there are only a very limited number of possibilities to consider (especially if we confine ourselves to unramified primes, as in fact we can), we can reasonably hope to have product relations between $\zeta_K(s)$ and the various $\zeta_L(s)$ including $\zeta_{\mathbf{Q}}(s)$. All this is more easily understood from the example to be found in §12. For the general theory, see [FT], Chapter VIII.7.

15 Analytic continuation and the functional equation

In this section μ^+ will denote the Haar measure on V_k, normalized so that $\mu^+(V_k/k) = 1$, and μ will be a Haar measure on J_k; we shall defer for as long as possible introducing the factor which comes from the particular normalization of μ which we chose in §A1.3. We now define a Tate zeta function; we shall see later how it relates to Hecke's zeta function. Let χ be a continuous character on J_k which is trivial on k^*; we shall call such a character a **Tate** character. Let $f : V_k \to \mathbf{C}$ be a function so well behaved that all the formal manipulations which follow are valid. By analogy with the Fourier transform on J_k we define

$$\zeta(f, \chi, s) = \int_{J_k} f(\alpha)\chi(\alpha)\|\alpha\|^s d\mu, \qquad (80)$$

where we require $f(\alpha)$ to die away so rapidly as $\|\alpha\| \to \infty$ that the integral converges absolutely in some right hand half-plane $\Re s > \sigma_0$. This formula defines an analytic function of s in this half-plane; our main task is to continue it to the entire s-plane. For this it is crucial that χ is trivial on k^*.

We divide J_k into two parts: $J^>$ on which $\|\alpha\| > 1$ and $J^<$ on which $\|\alpha\| < 1$; we can ignore the part with $\|\alpha\| = 1$ because it has measure 0. The integral over $J^>$ is absolutely convergent and defines an analytic function for all s, since these properties hold for $\Re s > \sigma_0$; it is the integral over $J^<$ which we have to worry about. Let $\mathcal{S}$ be a fundamental domain for the action of k^* on $J^<$ and write $J^< = \cup \xi \mathcal{S}$ where ξ runs through the elements of k^*; then

$$\int_{J^<} f(\alpha)\chi(\alpha)\|\alpha\|^s d\mu = \sum_{\xi \text{ in } k^*} \int_{\mathcal{S}} f(\xi\alpha)\chi(\alpha)\|\alpha\|^s d\mu$$

$$= \int_{\mathcal{S}} \left\{ \sum_{\xi \text{ in } k} f(\alpha\xi) \right\} \chi(\alpha)\|\alpha\|^s d\mu - f(0) \int_{\mathcal{S}} \chi(\alpha)\|\alpha\|^s d\mu.$$

In the first term on the right we apply the corollary to Theorem A6 (the Poisson Summation Formula); since $\mu^+(V_k/k) = 1$, we can replace the term in curly brackets by

$$\|\alpha\|^{-1} \sum_{\xi \text{ in } k} \hat{f}(\alpha^{-1}\xi).$$

Write β^{-1} for α and note that in an obvious notation $d\mu(\beta) = d\mu(\alpha)$ and

$\mathcal{S}^{-1}$ is a fundamental domain for the action of k^* on $J^>$. Thus

$$\int_{\mathcal{S}} \left\{ \sum_{\xi \text{ in } k} f(\alpha\xi) \right\} \chi(\alpha)\|\alpha\|^s d\mu = \int_{\mathcal{S}} \left\{ \sum_{\xi \text{ in } k} \hat{f}(\alpha^{-1}\xi) \right\} \chi(\alpha)\|\alpha\|^{s-1} d\mu$$

$$= \int_{\mathcal{S}^{-1}} \left\{ \sum_{\xi \text{ in } k} \hat{f}(\beta\xi) \right\} \chi(\beta^{-1})\|\beta\|^{1-s} d\mu.$$

Defining $\hat{\chi}$ by $\hat{\chi}(\beta) = \chi(\beta^{-1})$ and interchanging integration and summation,

$$\left. \begin{aligned} \int_{J^<} f(\alpha)\chi(\alpha)\|\alpha\|^s d\mu + f(0) \int_{\mathcal{S}} \chi(\alpha)\|\alpha\|^s d\mu \\ = \int_{J^>} \hat{f}(\beta)\hat{\chi}(\beta)\|\beta\|^{1-s} d\mu + \hat{f}(0) \int_{\mathcal{S}^{-1}} \hat{\chi}(\beta)\|\beta\|^{1-s} d\mu. \end{aligned} \right\} \tag{81}$$

Let us further assume that $\hat{f}(\beta)$ dies away so rapidly as $\|\beta\| \to \infty$ that $\zeta(\hat{f}, \hat{\chi}, s)$ also exists for $\Re s$ large and positive; then the first integral on the right exists for all s.

We must now look at the second term on each side of (81). Suppose first that χ is not trivial on J_k^1, and choose γ in J_k^1 with $\chi(\gamma) \neq 1$. Writing $\alpha\gamma$ for α, we obtain

$$\int_{\mathcal{S}} \chi(\alpha)\|\alpha\|^s d\mu = \chi(\gamma) \int_{\mathcal{S}} \chi(\alpha)\|\alpha\|^s d\mu;$$

so this expression must vanish. Hence in this case the second term on the left in (81) vanishes, and the same is true of the second term on the right. On the other hand, if χ is trivial on J_k^1 then the value of $\chi(\alpha)$ depends only on $\|\alpha\|$ and hence $\chi(\alpha) = \|\alpha\|^c$ for some pure imaginary c, since these are the only characters on $J_k/J_k^1 \approx \mathbf{R}^+$. Here we can set $c = 0$, which is equivalent to absorbing c into s. Using the decomposition $J_k = J_k^1 \times (0, \infty)$ by which we defined the induced measure on J_k^1, we obtain

$$\int_{\mathcal{S}} \|\alpha\|^s d\mu = \mu^1(J_k^1/k^*) \int_0^1 t^s \frac{dt}{t} = \frac{\mu^1(J_k^1/k^*)}{s}$$

where μ^1 is the induced measure on J_k^1. The other similar integral is equal to $\mu^1(J_k^1/k^*)/(s - 1)$. Thus we finally obtain

Theorem 28 *Assume that if χ is trivial on J_k^1 then it is normalized so as*

to be trivial on J_k. Then under suitable conditions on f,

$$
\left.
\begin{aligned}
\zeta(f,\chi,s) = \int_{J^>} \{f(\alpha)\chi(\alpha)\|\alpha\|^s + \hat{f}(\alpha)\hat{\chi}(\alpha)\|\alpha\|^{1-s}\}d\mu \\
+ \left(\frac{\hat{f}(0)}{s-1} - \frac{f(0)}{s}\right)\mu^1(J_k^1/k^*)
\end{aligned}
\right\}
\tag{82}
$$

where the last term only occurs if $\chi(\alpha)$ is trivial. The right hand side is analytic in the entire s-plane except for the possible poles at $s = 1$ and $s = 0$ arising from the last term on the right. Moreover

$$
\zeta(f,\chi,s) = \zeta(\hat{f},\hat{\chi},1-s). \tag{83}
$$

The last sentence follows immediately from the main formula, bearing in mind that we have identified $\widehat{V_k}$ with V_k, so that the transform of $\hat{f}(\alpha)$ viewed as a function on V_k is $f(-\alpha)$.

Note that (82) is homogeneous in the Haar measure μ on J_k, so that we have not yet committed ourselves to a particular normalization of μ. Subject to this, the value of $\mu^1(J_k^1/k^*)$ can be read off from Theorem 23. The reader will have noticed that we have not used the fact that χ is a character, but only that it is multiplicative; so we could have taken χ to be any quasi-character $J_k \to \mathbf{C}^*$ trivial on k^*. But this apparent extra generality is spurious. For since J_k^1/k^* is compact, χ induces a character on J_k^1; and since the kernel of the map $\alpha \mapsto \|\alpha\|$ is J_k^1, we must have $|\chi(\alpha)| = \|\alpha\|^c$ for some real c. Now replacing χ by the character $\alpha \mapsto \|\alpha\|^{-c}\chi(\alpha)$ is equivalent to replacing s by $s - c$.

To make use of these results, we need to evaluate the integrals which define $\hat{f}$ and ζ; and in any case we need to ensure that our hypotheses on f and χ can be satisfied. It is natural to choose f and χ to be products of functions defined on the factors k_v and k_v^* respectively:

$$
f(\xi) = \prod_v f_v(\xi_v), \quad \chi(\alpha) = \prod_v \chi_v(\alpha_v).
$$

The easy way to ensure that these products converge is to require for almost all v that $f_v = 1$ on $\mathfrak{o}_v$ and $\chi_v = 1$ on $\mathfrak{o}_v^*$. If we also require f_v to vanish outside $\mathfrak{o}_v$ for almost all v, then

$$
\hat{f}(\xi) = \int_{V_k} f(\eta)c(\xi\eta)d\mu = \prod \widehat{f_v}(\xi_v);
$$

for the integral over V_k is equal to the product of the integrals over the k_v and $c(\alpha) = \prod c_v(\alpha_v)$.

A similar argument gives the corresponding result

$$\zeta(f,\chi,s) = \prod_v \int_{k_v^*} f_v(\alpha_v)\chi_v(\alpha_v)\|\alpha_v\|_v^s d\mu_v, \tag{84}$$

where now the measure on J_k is the product of the measures on the k_v^*, as in Theorem 23. For if v is $\mathfrak{p}$ and $f_\mathfrak{p}$ vanishes outside $\mathfrak{o}_\mathfrak{p}$, the corresponding factor is

$$\zeta_\mathfrak{p}(f_\mathfrak{p},\chi_\mathfrak{p},s) = \sum_{n=0}^{\infty} \int_{\mathfrak{p}^n\backslash\mathfrak{p}^{n+1}} f_\mathfrak{p}(\alpha_\mathfrak{p})\chi_\mathfrak{p}(\alpha_\mathfrak{p})\|\alpha_\mathfrak{p}\|_\mathfrak{p}^s d\mu_\mathfrak{p}. \tag{85}$$

Here the n-th term is the product of $(\mathrm{Norm}\,\mathfrak{p})^{-ns}$ and a coefficient independent of s. If the formal product of these series is absolutely convergent and so are the remaining integrals in (84), then the non-zero terms of the product are precisely the ones we need for an integral over J_k^1. Here we treat the product over the non-Archimedean valuations as a Dirichlet series and therefore only retain the corresponding terms in the product; this is the same convention which we used to expand the Euler product in (71).

We now evaluate the local factors for those primes $\mathfrak{p}$ for which $\chi_\mathfrak{p} = 1$ on $\mathfrak{o}_\mathfrak{p}^*$. We choose

$$f_\mathfrak{p}(\xi_\mathfrak{p}) = \begin{cases} 1 & \text{for } \xi_\mathfrak{p} \text{ in } \mathfrak{o}_\mathfrak{p}, \\ 0 & \text{otherwise.} \end{cases} \tag{86}$$

If $\mathfrak{d}_\mathfrak{p}$ denotes the local different for $k/\mathbf{Q}$ at $\mathfrak{p}$ then as in §A2.2

$$\widehat{f}_\mathfrak{p}(\eta_\mathfrak{p}) = \int_{\mathfrak{o}_\mathfrak{p}} \exp(2\pi i \iota_\mathfrak{p} \mathrm{Tr}(\xi_\mathfrak{p}\eta_\mathfrak{p}))d\mu_\mathfrak{p}^+ = \begin{cases} (\mathrm{Norm}\,\mathfrak{d}_\mathfrak{p})^{-1/2} & \text{for } \eta_\mathfrak{p} \text{ in } \mathfrak{d}_\mathfrak{p}^{-1}, \\ 0 & \text{otherwise,} \end{cases}$$

because the integrand is a character on $\mathfrak{o}_\mathfrak{p}$ which is trivial if and only if $\eta_\mathfrak{p}$ is in $\mathfrak{d}_\mathfrak{p}^{-1}$. By hypothesis $\chi_\mathfrak{p}(\alpha_\mathfrak{p})$ depends only on $\|\alpha_\mathfrak{p}\|_\mathfrak{p}$ and must therefore have the form

$$\chi_\mathfrak{p}(\alpha_\mathfrak{p}) = \|\alpha_\mathfrak{p}\|_\mathfrak{p}^{s_1}$$

for s_1 pure imaginary; here s_1 may depend on $\mathfrak{p}$. Using (85), the corresponding local factor of the zeta function $\zeta(f,\chi,s)$ is then

$$\sum_0^{\infty} (\chi_\mathfrak{p}(\pi))^n (\mathrm{Norm}\,\mathfrak{p}^n)^{-s} = (1 - \chi_\mathfrak{p}(\pi)(\mathrm{Norm}\,\mathfrak{p})^{-s})^{-1}$$

where π satisfies $\mathfrak{p}\|\pi$. Similarly, if $\mathfrak{p}^r\|\mathfrak{d}_\mathfrak{p}$ the local factor of $\zeta(\widehat{f},\widehat{\chi},s)$ is

$$(\mathrm{Norm}\,\mathfrak{d}_\mathfrak{p})^{s-1/2}\chi(\pi^r)(1 - \widehat{\chi_\mathfrak{p}}(\pi)(\mathrm{Norm}\,\mathfrak{p})^{-s})^{-1}.$$

Let S be a finite set of places which includes the Archimedean places and all the primes $\mathfrak{p}$ excluded from the discussion above. Write $\chi^\flat(\mathfrak{p}) = \chi_\mathfrak{p}(\pi)$ for all $\mathfrak{p}$ outside $\mathcal{S}$ and use multiplicativity to extend the definition of $\chi^\flat$ to $I^{\mathcal{S}}$, the group of those fractional ideals in I_k whose factorizations do not involve any prime ideal in $\mathcal{S}$. We shall see in Lemma 34 below that the characters $\chi^\flat$ obtained in this way are precisely the Hecke Grössencharakters, and $L(s, \chi^\flat)$ below is the Hecke L-series. Thus up to finitely many factors corresponding to the places in $\mathcal{S}$,

$$\zeta(f, \chi, s) = \prod (1 - \chi^\flat(\mathfrak{p})(\mathrm{Norm}\,\mathfrak{p})^{-s})^{-1},$$

where by analogy with (78) it is natural to call the right hand side $L(s, \chi^\flat)$. It follows from (83) and the expressions above for the local factors that

$$\frac{L(1 - s, \widehat{\chi^\flat})}{L(s, \chi^\flat)} = \chi^\flat(\mathfrak{d}_*^{-1})(\mathrm{Norm}\,\mathfrak{d}_*)^{s-1/2} \prod_{\mathcal{S}} \frac{\zeta_v(f_v, \chi_v, s)}{\zeta_v(\widehat{f}_v, \widehat{\chi}_v, 1 - s)} \qquad (87)$$

where $\mathfrak{d}_*$ is the product of the $\mathfrak{d}_\mathfrak{p}$ for $\mathfrak{p}$ not in $\mathcal{S}$. In other words, we get a functional equation for $L(s, \chi^\flat)$, provided that the local characters χ_v have been chosen so that $\chi = \prod \chi_v$ is trivial on k^*. To satisfy this condition, we need to see what local characters χ_v are available.

Suppose first that v comes from a finite prime $\mathfrak{p}$. The $1 + \mathfrak{p}^m$ are a decreasing sequence of small subgroups, in the sense explained on page 126; so $\chi_\mathfrak{p}$ is trivial on those which have $m \geq m_\mathfrak{p}$, where we take $m_\mathfrak{p}$ to be as small as possible. We call $\mathfrak{p}^{m_\mathfrak{p}}$ the **conductor** of the restriction of $\chi_\mathfrak{p}$ to $\mathfrak{o}_\mathfrak{p}^*$; that restriction is induced by a character on the finite group $\mathfrak{o}_\mathfrak{p}^*/(1 + \mathfrak{p}^{m_\mathfrak{p}})$, and we are free to select any such character as $\chi_\mathfrak{p}$. Now fix an element π in k^* with $\mathfrak{p} \| \pi$ and write $\alpha_\mathfrak{p} = \pi^\nu \beta_\mathfrak{p}$ where $\beta_\mathfrak{p}$ is in $\mathfrak{o}_\mathfrak{p}^*$; if we define s_v by $\chi_\mathfrak{p}(\pi) = \|\pi\|_\mathfrak{p}^{s_v}$ then the most general $\chi_\mathfrak{p}$ is given by

$$\chi_\mathfrak{p}(\alpha_\mathfrak{p}) = \chi_\mathfrak{p}(\beta_\mathfrak{p}) \|\alpha_\mathfrak{p}\|_\mathfrak{p}^{s_v}.$$

Here s_v is really an element of $i\mathbf{R} \bmod (2\pi i / \log \|\pi\|_\mathfrak{p})$ rather than of $i\mathbf{R}$. Next suppose that v is real. Since the characters on $(0, \infty)$ are just the $x \to x^{s_v}$ for some pure imaginary s_v, there are just two classes of characters:

$$\chi_v(x) = |x|^{s_v} \quad \text{or} \quad \chi_v(x) = |x|^{s_v} \mathrm{sign}\, x. \qquad (88)$$

Finally, suppose that v is complex. Let $z = re^{i\theta}$; then by considering the restrictions of χ_v to $\theta = 0$ and to $r = 1$ we obtain

$$\chi_v(z) = r^{s_v} e^{in_v\theta} \qquad (89)$$

for some pure imaginary s_v and some integer n_v. The effect of increasing all the s_v by the same constant c is to multiply $\chi(\alpha)$ by $\|\alpha\|^c$, which is equivalent to a translation on s; but in general there is no obviously natural way of choosing c. Also, replacing χ by χ^{-1} reverses the signs of all the s_v.

We have still to ensure that $\chi = \prod \chi_v$ is trivial on k^*; we achieve this by forcing χ to be trivial successively on roots of unity, on $\mathfrak{o}_k^*$ modulo roots of unity, and on $k^*/\mathfrak{o}_k^*$. First choose the presence or absence of the factor $\operatorname{sign} x$ in (88) for each real valuation and the value of n in (89) for each complex valuation. Choose also a finite set $\mathcal{S}$ of non-Archimedean valuations (which will be the ones for which $\chi_\mathfrak{p}$ is non-trivial on $\mathfrak{o}_\mathfrak{p}^*$) and for each of them choose a character $\chi_\mathfrak{p}$ on $\mathfrak{o}_\mathfrak{p}^*$, subject only to the condition that $\chi(\omega) = 1$ if ω generates the group of roots of unity in k^*. The **conductor** of χ, denoted by $\mathfrak{f}$, will be the product of the local conductors $\mathfrak{p}^{m_\mathfrak{p}}$ defined above. We shall show that for each set of such choices there are exactly h admissible characters χ, where h is the class number of k.

The values of s_v for the Archimedean valuations are determined up to the addition of a common constant by the condition that χ is trivial on a base for $\mathfrak{o}_k^*$ modulo roots of unity; for this imposes $r_1 + r_2 - 1$ linear conditions on these s_v whose determinant (with some abuse of language) is the regulator. These equations have the form $\sum a_{\mu\nu} s_{v_\nu} = ib_\mu$ where the $a_{\mu\nu}, b_\mu$ are real; so we can take the s_v to be pure imaginary. In view of the remarks after (89) we can assume a fixed choice of these values. Now denote by J_k^{triv} the group of idèles $\beta = \prod \beta_v$ such that $\beta_\mathfrak{p}$ is in $\mathfrak{o}_\mathfrak{p}^*$ for each $\mathfrak{p}$; thus J^{triv} is both the kernel of the map $J_k \to I_k$ and the set of idèles for which the choices above have already determined the value of χ. There is an exact sequence

$$0 \to \mathfrak{o}_k^* \to k^* \times J_k^{\mathrm{triv}} \to J_k \to \mathcal{C}_k \to 0, \tag{90}$$

in which the middle map is $\alpha \times \beta \mapsto \alpha^{-1}\beta$ and the one following it is $J_k \to I_k \to \mathcal{C}_k$. Define χ on the image of ϕ by $\chi(\phi(\alpha \times \beta)) = \chi(\beta)$ where ϕ is the middle map in (90); χ is well-defined because β is determined up to multiplication by an element of $\mathfrak{o}_k^*$ and we have already arranged that χ is trivial on $\mathfrak{o}_k^*$. Using Theorem A1, we see that χ can be extended to a character on J_k in just h ways; and it is easy to see that these are precisely the characters on J_k which extend the restriction of χ to J^{triv} and are trivial on k^*.

To express the condition which defines a Hecke Grössencharakter ψ for an algebraic number field k requires an integral ideal $\mathfrak{f}$ of k, a pure imaginary number s_v for each infinite place v of k and a rational integer n_v for each complex place v. Let $\mathcal{S}$ be the set of prime ideals which divide $\mathfrak{f}$ and denote by $I^\mathcal{S}$ the set of fractional ideals in k whose prime factorization involves no ideal in $\mathcal{S}$. A character ψ on $I^{\mathcal{S}}$ is called a **Hecke Grössencharakter** if

$$\psi((\alpha)) = \left\{ \prod |\sigma_v \alpha|^{-s_v} \right\} \left\{ {\prod}'(\sigma_v \alpha / |\sigma_v \alpha|)^{-n_v} \right\} \qquad (91)$$

whenever $\alpha \equiv 1 \bmod \mathfrak{f}$ and $\sigma_v \alpha > 0$ for all real embeddings $\sigma_v : k \to \mathbf{R}$. In (91) σ_v is the embedding $k \to \mathbf{C}$, the first product is taken over all infinite places and the second over all complex infinite places. For given $\mathfrak{f}$ we are free to choose the n_v, but there are considerable constraints on the s_v arising from the fact that $\psi((\alpha)) = 1$ whenever α is a unit.

Lemma 34 *The characters χ^b derived from Tate characters χ are the Hecke Grössencharakters.*

Proof Let χ be a Tate character with conductor $\mathfrak{f}$, and let n_v, s_v be as in (88) and (89). If α is as above then $\chi_{\mathfrak{p}}(\alpha) = 1$ for $\mathfrak{p}$ in $\mathcal{S}$, so $1 = \chi(\alpha) = \prod \chi_v(\alpha)$ implies

$$\chi^b((\alpha)) = \prod_{\mathfrak{p} \text{ not in } \mathcal{S}} \chi_{\mathfrak{p}}(\alpha) = \left\{ \prod \chi_v(\alpha) \right\}^{-1}$$

where the right hand product is over all Archimedean v; and this is (91).

Conversely, suppose that ψ satisfies (91). We shall require χ to have conductor $\mathfrak{f}$, so that $\chi_{\mathfrak{p}}$ for $\mathfrak{p}$ not in $\mathcal{S}$ is completely specified by the requirement $\chi_{\mathfrak{p}}(\pi) = \psi(\mathfrak{p})$. Since χ^b only depends on these $\chi_{\mathfrak{p}}(\pi)$, this ensures that $\chi^b = \psi$ if we can construct χ at all. If v is Archimedean then χ_v is specified by (88) and (89) where n_v, s_v are as in (91), except that we have still the choice between the two alternatives in (88). By the argument of the previous paragraph we have ensured that $\chi(\alpha) = 1$ for all $\alpha \equiv 1 \bmod \mathfrak{f}$ such that α is positive at all real places. If merely $\alpha \equiv 1 \bmod \mathfrak{f}$ then $\chi(\alpha^2) = 1$; since

$$\{\alpha \equiv 1 \bmod \mathfrak{f}\} / \{\alpha \equiv 1 \bmod \mathfrak{f} \text{ and } \alpha > 0 \text{ at all real places}\} \approx \{\pm 1\}^{r_1},$$

there is just one way of making the r_1 choices (88) so that $\chi(\alpha) = 1$ for all $\alpha \equiv 1 \bmod \mathfrak{f}$. Similarly there is then just one way of choosing $\chi_{\mathfrak{p}}$ on $\mathfrak{o}_{\mathfrak{p}}^*$ for each $\mathfrak{p}$ in $\mathcal{S}$ so that $\chi(\alpha) = 1$ for all α which are units at $\mathfrak{p}$ for each such $\mathfrak{p}$, and finally just one way of choosing the $\chi_{\mathfrak{p}}(\pi)$ for $\mathfrak{p}$ in $\mathcal{S}$ to ensure that $\chi(\alpha) = 1$ for all α in k^*.

All that is left to do is some tidying up. It follows from (87) that the quotient

$$\zeta_v(\widehat{f_v}, \widehat{\chi_v}, 1 - s) / \zeta_v(f_v, \chi_v, s)$$

does not depend on the choice of f_v. This can be proved directly, because it is equivalent to

$$\zeta_v(f_v, \chi_v, s)\zeta_v(\widehat{g}_v, \widehat{\chi}_v, 1 - s) = \zeta_v(g_v, \chi_v, s)\zeta_v(\widehat{f}_v, \widehat{\chi}_v, 1 - s)$$

for any admissible functions f_v, g_v on k_v. The left hand side here is

$$\int_{k_v^*} \int_{k_v^*} f_v(\alpha)\widehat{g}_v(\beta)\chi_v(\alpha\beta^{-1})\|\alpha\beta^{-1}\|_v^s \|\beta\|_v \, d\mu_\alpha d\mu_\beta$$

$$= \int_{k_v^*} \left\{ \int_{k_v^*} f_v(\alpha)\widehat{g}_v(\alpha\gamma)\|\alpha\|_v \, d\mu_\alpha \right\} \chi_v(\gamma^{-1})\|\gamma\|_v^{1-s} \, d\mu_\gamma$$

on writing $\beta = \alpha\gamma$. Since $\|\alpha\|_v d\mu_\alpha = \lambda_v d\mu_\alpha^+$ for a constant λ_v whose value is unimportant, the expression in curly brackets is

$$\lambda_v^2 \int_{k_v} f_v(\alpha) \left\{ \int_{k_v} g_v(\delta)c_v(\alpha\gamma\delta) \, d\mu_\delta^+ \right\} d\mu_\alpha^+$$

where c_v is the additive character which appears in the Fourier transform. This last expression is clearly symmetric in f_v and g_v.

Nevertheless, we should choose these f_v to make everything as simple as possible. If v is real the best choice for f_v is

$$f_v(x) = \begin{cases} \exp(-\pi x^v) & \text{if} \quad \chi_v(x) = |x|^{s_v}, \\ x\exp(-\pi x^2) & \text{if} \quad \chi_v(x) = |x|^{s_v}\operatorname{sign} x, \end{cases} \tag{92}$$

where the reader is warned that π denotes $3 \cdot 14159\ldots$ when discussing infinite places. Now $\widehat{f}_v(x) = f_v(x)$ in the first case and $\widehat{f}_v(x) = if_v(x)$ in the second. Indeed these are just the identities

$$\int_{-\infty}^{\infty} \exp(-\pi y^2 + 2\pi ixy)dy = \exp(-\pi x^2), \tag{93}$$

$$\int_{-\infty}^{\infty} y\exp(-\pi y^2 + 2\pi ixy)dy = ix\exp(-\pi x^2), \tag{94}$$

where (93) is standard and (94) is obtained by differentiating with respect to x. The corresponding factor $\zeta_v(f_v, \chi_v, s)$ reduces to the standard integral for the Gamma function; its value is $\pi^{-(s+s_v)/2}\Gamma(\frac{1}{2}(s + s_v))$ in the first case and $\pi^{-(s+s_v+1)/2}\Gamma(\frac{1}{2}(s + s_v + 1))$ in the second. Similarly the value of $\zeta_v(\widehat{f}_v, \widehat{\chi}_v, 1 - s)$ is obtained by writing $1 - s$ for s and reversing the sign of s_v, and also multiplying by i in the second case.

If v is complex then we write $z = x + iy$ and in an obvious notation take

$$f_{v,n} = \begin{cases} (x - iy)^{|n|} \exp(-2\pi(x^2 + y^2)) & \text{if} \quad n \geqslant 0, \\ (x + iy)^{|n|} \exp(-2\pi(x^2 + y^2)) & \text{if} \quad n \leqslant 0. \end{cases}$$

I claim that this gives $\widehat{f_{v,n}}(z) = i^{|n|} f_{v,-n}(z)$. The minus sign on the right ensures that $\widehat{\chi}(re^{i\theta}) = r^{-2s_v}e^{-in\theta}$. For $n = 0$ the assertion is that

$$\int_{-\infty}^{\infty}\int_{-\infty}^{\infty} \exp(-2\pi(u^2 + v^2) + 4\pi i(xu - yv))du\,dv = \tfrac{1}{2}\exp(-2\pi(x^2 + y^2)),$$

which is just the product of (93) with itself. The general case is now obtained by applying the operator $\left(\frac{\partial}{\partial x} + i\frac{\partial}{\partial y}\right)^n$ if $n \geqslant 0$, or $\left(\frac{\partial}{\partial x} - i\frac{\partial}{\partial y}\right)^{-n}$ if $n \leqslant 0$. The factor $\zeta_v(f_v, \chi_v, s)$ is $(2\pi)^{1-s-s_v+|n|/2}\Gamma(s + s_v + \tfrac{1}{2}|n|)$. The value of $\zeta_v(\widehat{f_v}, \widehat{\chi_v}, 1 - s)$ is again obtained from this by writing $1 - s$ for s and reversing the sign of s_v.

The remaining cases are when v is non-Archimedean and $\mathfrak{p}|\mathfrak{f}$. Now we can no longer define $f_\mathfrak{p}$ by (86) because $\zeta_\mathfrak{p}(f_\mathfrak{p}, \chi_\mathfrak{p}, s)$ would vanish if we did so; there is no obviously best choice for $f_\mathfrak{p}$, and we shall take

$$f_\mathfrak{p}(x) = \begin{cases} c_1(x) = \exp(2\pi i \operatorname{Tr}_{k_\mathfrak{p}/\mathbf{Q}_p}(x)) & \text{for } x \text{ in } \mathfrak{d}_\mathfrak{p}^{-1}\mathfrak{f}_\mathfrak{p}^{-1}, \\ 0 & \text{otherwise} \end{cases}$$

in the notation of Lemma A6, where $\mathfrak{f}_\mathfrak{p}$ is the conductor of $\chi_\mathfrak{p}$ as on page 88. Thus

$$\widehat{f_\mathfrak{p}}(y) = \int_{\mathfrak{d}_\mathfrak{p}^{-1}\mathfrak{f}_\mathfrak{p}^{-1}} c_{1-y}(x)\,d\mu_x^+ = \begin{cases} (\operatorname{Norm}\mathfrak{d}_\mathfrak{p})^{1/2}\operatorname{Norm}\mathfrak{f}_\mathfrak{p} & \text{if } y \equiv 1 \bmod \mathfrak{f}_\mathfrak{p}, \\ 0 & \text{otherwise,} \end{cases}$$

by Lemma A5. If for convenience we define m, r by $\mathfrak{p}^m \| \mathfrak{f}_\mathfrak{p}$, $\mathfrak{p}^r \| \mathfrak{d}_\mathfrak{p}$, then

$$\zeta_\mathfrak{p}(f_\mathfrak{p}, \chi_\mathfrak{p}, s) = \sum_{n=-m-r}^{\infty} \|\mathfrak{p}\|^{ns} \int_{\mathfrak{p}^n \setminus \mathfrak{p}^{n+1}} c_1(x)\chi_\mathfrak{p}(x)\,d\mu_x.$$

I claim that every term in this sum vanishes except the first.

Suppose first that $n \geqslant -r$, so that $c_1(x) = 1$ on $\mathfrak{p}^n \setminus \mathfrak{p}^{n+1}$. Writing $x = \pi^n y$, the integral becomes

$$\int_{\mathfrak{p}^n \setminus \mathfrak{p}^{n+1}} \chi_\mathfrak{p}(x)\,d\mu_x = \chi_\mathfrak{p}(\pi^n)\int_{\mathfrak{o}_\mathfrak{p}^*} \chi_\mathfrak{p}(y)\,d\mu_y = 0$$

because the restriction of $\chi_\mathfrak{p}$ to $\mathfrak{o}_\mathfrak{p}^*$ is non-trivial.

Now suppose that $-m - r < n < -r$. We write $\mathfrak{p}^n \setminus \mathfrak{p}^{n+1}$ as a union of sets $x_0 + \mathfrak{p}^{-r} = x_0(1 + \mathfrak{p}^{-n-r})$. On such a set $c_1(x) = c_1(x_0)$, so

$$\int_{\mathfrak{p}^n \setminus \mathfrak{p}^{n+1}} c_1(x)\chi_\mathfrak{p}(x)\,d\mu_x = \sum_{x_0} c_1(x_0)\chi_\mathfrak{p}(x_0)\int_{1+\mathfrak{p}^{-n-r}} \chi_\mathfrak{p}(y)\,d\mu_y;$$

but $0 < -n - r < m$, so $\chi_\mathfrak{p}$ is a non-trivial character on $1 + \mathfrak{p}^{-n-r}$ and each integral on the right vanishes. Thus finally

$$\frac{\zeta_{\mathfrak{p}}(f_{\mathfrak{p}}, \chi_{\mathfrak{p}}, s)}{\zeta_{\mathfrak{p}}(\widehat{f_{\mathfrak{p}}}, \widehat{\chi_{\mathfrak{p}}}, 1 - s)} = (\text{Norm}\, \mathfrak{f}_{\mathfrak{p}}\mathfrak{d}_{\mathfrak{p}})^{s-1/2}\chi_{\mathfrak{p}}(\pi)^{-m-r}W_{\mathfrak{p}}(\chi_{\mathfrak{p}}), \qquad (95)$$

where writing $x\pi^{-m-r}$ for x in the integral gives

$$W_{\mathfrak{p}}(\chi_{\mathfrak{p}}) = (\text{Norm}\, \mathfrak{f}_{\mathfrak{p}})^{1/2}\int_{\mathfrak{o}_{\mathfrak{p}}^*} c_1(x\pi^{-m-r})\chi_{\mathfrak{p}}(x)\, d\mu_x. \qquad (96)$$

If, copying the earlier argument, we write $x = \epsilon(1+y)$ where ϵ runs through a set of representatives for $\mathfrak{o}_{\mathfrak{p}}^*/(1 + \mathfrak{p}^m)$ and y is in $\mathfrak{p}^m$, then we obtain

$$W_{\mathfrak{p}}(\chi_{\mathfrak{p}}) = (\text{Norm}\, \mathfrak{f}_{\mathfrak{p}})^{-1/2}\sum c_1(\epsilon\pi^{-m-r})\chi_{\mathfrak{p}}(\epsilon)$$

which can be described as a generalized local Gauss sum. The canonical property

$$|W_{\mathfrak{p}}(\chi_{\mathfrak{p}})| = 1 \qquad (97)$$

can be proved as in §13.1, or we can argue as follows. On the one hand, the factor associated with $\mathfrak{p}$ on the right of (87) is independent of the choice of $f_{\mathfrak{p}}$. If instead we use $g_{\mathfrak{p}} = \widehat{f_{\mathfrak{p}}}$ then $\widehat{g_{\mathfrak{p}}}(x) = f_{\mathfrak{p}}(-x)$; so this factor is equal to

$$\frac{\zeta_{\mathfrak{p}}(g_{\mathfrak{p}}, \chi_{\mathfrak{p}}, s)}{\zeta_{\mathfrak{p}}(\widehat{f_{\mathfrak{p}}}, \widehat{\chi_{\mathfrak{p}}}, 1 - s)} = \chi_{\mathfrak{p}}(-1)\left\{\frac{\zeta_{\mathfrak{p}}(f_{\mathfrak{p}}, \widehat{\chi_{\mathfrak{p}}}, 1 - s)}{\zeta_{\mathfrak{p}}(\widehat{f_{\mathfrak{p}}}, \chi_{\mathfrak{p}}, s)}\right\}^{-1}$$

$$= \chi_{\mathfrak{p}}(-1)(\text{Norm}\,(\mathfrak{d}_{\mathfrak{p}}\mathfrak{f}_{\mathfrak{p}}))^{s-1/2}\widehat{\chi_{\mathfrak{p}}}(\pi^{m+r})/W_{\mathfrak{p}}(\widehat{\chi_{\mathfrak{p}}})$$

by (95), where the factor $\chi_{\mathfrak{p}}(-1)$ comes from the need to write $-x$ for x in the integral (96). Now set $s = \frac{1}{2}$ and write $\widehat{\chi_{\mathfrak{p}}}$ instead of $\chi_{\mathfrak{p}}$; comparing the last result with (95) we obtain

$$W_{\mathfrak{p}}(\chi_{\mathfrak{p}})W_{\mathfrak{p}}(\widehat{\chi_{\mathfrak{p}}}) = \chi_{\mathfrak{p}}(-1).$$

On the other hand, it follows from (96) on writing $-x$ for x that

$$\overline{W_{\mathfrak{p}}(\chi_{\mathfrak{p}})} = \chi_{\mathfrak{p}}(-1)W_{\mathfrak{p}}(\overline{\chi_{\mathfrak{p}}}) = \chi_{\mathfrak{p}}(-1)W_{\mathfrak{p}}(\widehat{\chi_{\mathfrak{p}}}).$$

These two results together give (97).

We can now finally write down the functional equation in full detail. We again define $\Lambda(s, \chi)$ by (79), though now $\Gamma(s, \chi)$ is the product of factors for each infinite place which are

$$\pi^{-(s+s_v)/2}\Gamma(\tfrac{1}{2}(s + s_v)) \text{ for the first case in (92),}$$

$$\pi^{-(s+s_v+1)/2}\Gamma(\tfrac{1}{2}(s + s_v + 1)) \text{ for the second case in (92),}$$

$$(2\pi)^{-s-s_v+|n|/2}\Gamma(s + s_v + \tfrac{1}{2}|n|) \text{ if } v \text{ is complex.}$$

Combining all our previous results now gives

$$\frac{\Lambda(1-s,\widehat{\chi^b})}{\Lambda(s,\chi^b)} = i^{-t} \prod_{\mathfrak{p} \text{ in } S} W_{\mathfrak{p}}(\chi_{\mathfrak{p}})$$

where t is the number of real places for which $\chi_v(x) = |x|^{s_v} \operatorname{sign} x$. $\qquad\square$

16 Density theorems

Euclid proved that there are an infinity of rational primes. Dirichlet proved that any congruence class $\bmod n$ which is prime to n contains an infinity of primes; and a much more general theorem of Čebotarev will be found below. On the other hand, it has not been proved that there are an infinity of n for which $n^2 + 1$ is prime, or for which n and $n + 2$ are both prime.

Most proofs that there are an infinity of primes having a given property depend on showing that the set of such primes has positive density. There are two standard measures of density even for sets of rational primes, and they both generalize to an arbitrary algebraic number field k. Let S be a set of prime ideals of k and S_0 the set of all prime ideals of k. The most common definition of the density of S is

$$\lim_{X\to\infty}(N(S,X)/N(S_0,X)) \qquad (98)$$

provided this limit exists; here $N(S,X)$ is the number of prime ideals $\mathfrak{p}$ in S which satisfy $\operatorname{Norm}\mathfrak{p} < X$ and similarly for $N(S_0,X)$. The prime number theorem for k (which we shall not prove) states that

$$N(S_0,X) = \frac{X}{\log X}(1 + o(1)).$$

The alternative measure is the **Dirichlet density**, which is defined as

$$\lim_{s\to 1} \left\{ \sum_{\mathfrak{p} \text{ in } S}(\operatorname{Norm}\mathfrak{p})^{-s} \Big/ \sum_{\mathfrak{p} \text{ in } S_0}(\operatorname{Norm}\mathfrak{p})^{-s} \right\} \qquad (99)$$

provided this limit exists. In order not to have to prove analytic continuation, it is usual to take the limit as s tends to 1 from above. The denominator in (99) is asymptotic to $-\log(s-1)$, as can easily be seen from (74) and (75).

Whether we work with (98) or (99), the density of S depends only on the first degree primes in S — that is, the prime ideals $\mathfrak{p}$ for which $\operatorname{Norm}\mathfrak{p}$ is a prime in $\mathbf{Q}$; for it is easy to show that the set of all primes in k which are not first degree has density zero.

If the limit (98) exists so does the limit (99) and they have the same value;

see Exercise 4.2. But (99) can exist even when (98) does not, though such cases are probably pathological. More to the point, (99) is usually much easier to prove than (98); for the natural line of attack on (98) requires a knowledge of the behaviour of the numerator of (99) on the whole line $\Re s = 1$. For applications, what is normally needed is to know that the density of S is strictly positive; but it does not matter which density one uses. Each of the three theorems in this section is stated and proved for Dirichlet density, but remains true if we replace Dirichlet density by density as defined by (98).

The standard way to prove a density theorem appears in its simplest form in Dirichlet's theorem on primes in arithmetic progression.

Theorem 29 *Let $m > 2$ and let c be a fixed integer prime to m; then the rational primes which satisfy $p \equiv c \bmod m$ have Dirichlet density $(\phi(m))^{-1}$ where ϕ is Euler's function.*

Proof Let $k = \mathbf{Q}(\sqrt[m]{1})$ and let χ be a character on $\mathrm{Gal}(k/\mathbf{Q})$; by the Corollary to Theorem 27, χ induces a character on $(\mathbf{Z}/m\mathbf{Z})^*$ which we shall also denote by χ. Suppose that χ is not principal. In the notation of (73)

$$L(s,\chi) = \sum_{1}^{\infty} S_n(n^{-s} - (n+1)^{-s}) \quad \text{where} \quad S_n = \sum_{1}^{n} \chi(r).$$

But $n^{-s} - (n+1)^{-s} = O(n^{-1-\sigma})$ where $\sigma = \Re s$, and $S_{\nu m} = 0$ for any integer ν by Lemma A5, so that S_n is bounded; so $L(s,\chi)$ is analytic in $\Re s > 0$ and in particular at $s = 1$. Again, if p has order f as an element of $(\mathbf{Z}/m\mathbf{Z})^*$ then $\chi(p)$ runs through the f-th roots of unity as χ varies, and $\chi(p)$ takes each value equally often; so

$$\prod_{\chi}(1 - \chi(p)p^{-s}) = (1 - p^{-fs})^{\phi(m)/f}.$$

Using the Corollary to Theorem 27 again, it follows that

$$\zeta_k(s) = \prod_{\chi} L(s,\chi) \tag{100}$$

up to finitely many factors corresponding to the primes dividing m. Here $\zeta_k(s)$ has a simple pole at $s = 1$ and so has $L(s,\chi_0)$ where χ_0 is the principal character; for $L(s,\chi_0)$ is $\zeta_{\mathbf{Q}}(s)$ up to finitely many factors. All the other factors on the right in (100) are analytic at $s = 1$, so (100) implies

$$L(1,\chi) \neq 0 \quad \text{for} \quad \chi \neq \chi_0.$$

Since $\log(1 - \chi(p)p^{-s}) = -\chi(p)p^{-s} + O(p^{-2\sigma})$ where $\sigma = \Re s$, we obtain

$$\log L(s, \chi) = \sum \chi(p)p^{-s} + O(1)$$

in $\Re s > 1$. Thus

$$\phi(m)\sum_{p \equiv c \bmod m} p^{-s} = \sum_p \sum_\chi \overline{\chi(c)}\chi(p)p^{-s}$$
$$= \sum_\chi \overline{\chi(c)} \log L(s, \chi) + O(1).$$

A similar argument gives

$$\log L(s, \chi_0) = \log \zeta_{\mathbf{Q}}(s) + O(1) = \sum_p p^{-s} + O(1).$$

Combining these two equations with (100) gives the theorem.   □

The same ideas work for an arbitrary algebraic number field k, but we have to appeal to much more sophisticated results. We now use the notation of §17. A weaker version of the following result will reappear as Theorem 35.

Theorem 30 *Let k be an algebraic number field and A/H a congruence divisor class group in k. Let C be a coset of H in A. Then the Dirichlet density of the prime ideals in C is N^{-1}, where $N = \mathrm{card}(A/H)$.*

Proof Let K be the abelian extension of k which is the class field for A/H in the sense of Theorem 32, and let χ run through the characters of A/H. If χ_0 is the principal character then $L(s, \chi_0)$ is equal to $\zeta_k(s)$ up to finitely many factors; so $L(s, \chi_0)$ has a simple pole at $s = 1$. Any other character χ satisfies $\chi((\alpha)) = 1$ whenever $\alpha \equiv 1 \bmod \mathfrak{f}$ and $\sigma_v \alpha > 0$ for all real places; so by Lemma 34 it comes from a Tate character which is clearly non-trivial. Hence $L(s, \chi)$ is analytic at $s = 1$, and using Theorem 37 instead of (100) we again obtain $L(1, \chi) \neq 0$. The rest of the proof follows that of the previous theorem.   □

The isomorphism between A/H and $\mathrm{Gal}(K/k)$ given by the Artin symbol implies that we can identify the set of prime ideals in C with the set of prime ideals $\mathfrak{p}$ for which the Artin symbol $\left(\frac{K/k}{\mathfrak{p}}\right)$ takes a given value. But this is only a special case of a more general result, the Čebotarev Density Theorem, for which we only need to assume that K/k is Galois and not necessarily that it is abelian.

Theorem 31 (Čebotarev) *Let K, k be algebraic number fields such that K is Galois over k, let σ be an element of $\mathrm{Gal}(K/k)$ and denote by $\langle \sigma \rangle$*

the conjugacy class of σ. *Let* S *be the set of prime ideals* $\mathfrak{p}$ *of* k *such that for every* $\mathfrak{P}$ *above* $\mathfrak{p}$ *the Frobenius element* $[\frac{K/k}{\mathfrak{P}}]$ *lies in* $\langle \sigma \rangle$. *Then* S *has Dirichlet density* $\mathrm{card}(\langle \sigma \rangle)/\mathrm{card}(\mathrm{Gal}(K/k))$.

Proof Let f be the order of σ and L the fixed field of σ, so that L is the splitting field of $\mathfrak{P}$ whenever $[\frac{K/k}{\mathfrak{P}}] = \sigma$. The Dirichlet density of the primes $\mathfrak{q}$ in L with $(\frac{K/L}{\mathfrak{q}}) = \sigma$ is f^{-1}, by the remarks after the proof of Theorem 30. If $\mathfrak{p}$ is in S its prime factors are the distinct $\tau\mathfrak{P}$ with τ in $\mathrm{Gal}(K/k)$; τ fixes $\mathfrak{P}$ if and only if τ is in the cyclic group generated by σ, and $[\frac{K/k}{\tau\mathfrak{P}}] = [\frac{K/k}{\mathfrak{P}}]$ if and only if τ is in $Z(\sigma)$, the centralizer of σ. Hence the number of distinct $\mathfrak{P}$ above $\mathfrak{p}$ with $[\frac{K/k}{\mathfrak{P}}] = \sigma$ is $f^{-1}\mathrm{card}(Z(\sigma))$; and the Dirichlet density of S is

$$\frac{f^{-1}}{f^{-1}\mathrm{card}(Z(\sigma))} = \frac{1}{\mathrm{card}(Z(\sigma))} = \frac{\mathrm{card}(\langle \sigma \rangle)}{\mathrm{card}(\mathrm{Gal}(K/k))}$$

because there is a one-one correspondence between the $\mathfrak{P}$ with $[\frac{K/k}{\mathfrak{P}}] = \sigma$ and the $\mathfrak{q}$ with $(\frac{K/L}{\mathfrak{q}}) = \sigma$, and $\mathrm{Norm}\,\mathfrak{q} = \mathrm{Norm}\,\mathfrak{p}$. $\qquad\square$

5

Class Field Theory

17 The classical theory

Let k be an algebraic number field and K a finite abelian extension of k. The objective of classical class field theory, which was largely achieved, was to describe the properties of K in terms of objects in k. The theory was first formulated in the 1890s, partly by Weber (following Kronecker) and partly by Hilbert (following Kummer); but one crucial component was only provided by Artin in 1927. The first proofs were given by Takagi in the 1920s; he used complicated group-theoretic arguments which we now know to belong to group cohomology — a subject which at that time had not been invented.

Let k be an algebraic number field with class number $h > 1$. It is straightforward to show that there are algebraic number fields $K \supset k$ such that every ideal in k becomes principal in K. Is there a canonical way of choosing K, and what additional properties will the extension K/k have? Hilbert conjectured that there is just one such field K with the following additional properties:

(i) $\mathfrak{d}_{K/k} = (1)$, so that the extension is unramified at all finite places;

(ii) the extension is also unramified at all infinite places, so that the places of K above a real infinite place of k are all real;

(iii) K is abelian over k with Galois group isomorphic to the ideal class group I_k of k.

He called such a field the **absolute class field** of k, and proved its existence in certain cases, for example when $h = 2$. This conjecture was reinforced by one of the key tools in Kummer's work on Fermat's Last Theorem: if k is an algebraic number field of class number h and K an abelian extension of k with $\mathfrak{d}_{K/k} = 1$ and $\mathrm{Gal}(K/k) \sim \mathbf{Z}/(p)$ where p is prime, then $p|h$ and there are just p ideal classes in k whose members become principal in K.

It had been shown by Kummer that the cyclotomic fields (the subfields of $\mathbf{Q}(\epsilon)$ where ϵ is a root of unity) were abelian over $\mathbf{Q}$ and that how a prime p factorized in such a field depended only on congruence properties of p. Kronecker conjectured the converse of this: that every abelian extension of $\mathbf{Q}$ is a cyclotomic field. A nearly complete proof was given by Weber, though it took a further generation before all the gaps in his argument were filled in. The result is however an easy consequence of the general theory, as well as being a major inspiration for its formulation.

The examples obtained in these ways led Weber to the general ideas of a **congruence divisor class group** and of a **class field**. To define the former, let $\mathfrak{m}$ be the formal product of an integral ideal in k and possibly some real infinite places in k; thus we can formally write $\mathfrak{m} = \prod \mathfrak{p}_\mu^{n_\mu}$ where some of the $\mathfrak{p}_\mu$ may be infinite real places. Denote by $A_\mathfrak{m}$ the subgroup of I_k consisting of the fractional ideals whose prime factorizations involve no prime ideal dividing $\mathfrak{m}$; here we only take account of the finite primes $\mathfrak{p}_\mu$ in $\mathfrak{m}$. Let $H_\mathfrak{m}^0$ be the subgroup of $A_\mathfrak{m}$ consisting of those principal ideals which can be written as (α) where $\alpha \equiv 1 \bmod \mathfrak{m}$ — that is, $\alpha \equiv 1 \bmod \mathfrak{p}_\mu^{n_\mu}$ in $k_{\mathfrak{p}_\mu}$ for each finite $\mathfrak{p}_\mu$ and $\sigma\alpha > 0$ if $\mathfrak{p}_\mu$ is a real infinite place and $\sigma: k \to \mathbf{R}$ is the corresponding map. In this way there corresponds to any $\mathfrak{m}$ the quotient group $A_\mathfrak{m}/H_\mathfrak{m}^0$, which is easily seen to be finite. In what follows, we shall denote by $H_\mathfrak{m}$ any subgroup of $A_\mathfrak{m}$ which contains $H_\mathfrak{m}^0$.

Suppose that $\mathfrak{m}|\mathfrak{n}$ in the obvious sense; then $A_\mathfrak{m} \supset A_\mathfrak{n}$ and $H_\mathfrak{m}^0 \supset H_\mathfrak{n}^0$. Moreover, $H_\mathfrak{n} = H_\mathfrak{m} \cap A_\mathfrak{n} \supset H_\mathfrak{n}^0$ for any $H_\mathfrak{m}$; and since each coset of $H_\mathfrak{m}^0$ in $A_\mathfrak{m}$ meets $A_\mathfrak{n}$ there is a canonical isomorphism

$$A_\mathfrak{n}/H_\mathfrak{n} \approx A_\mathfrak{m}/H_\mathfrak{m}.$$

Let $\mathfrak{m}_1, \mathfrak{m}_2$ be two such formal divisors with associated groups $H_{\mathfrak{m}_1}, H_{\mathfrak{m}_2}$. We shall say that these two groups are **equivalent** if for some common multiple $\mathfrak{m}$ of $\mathfrak{m}_1, \mathfrak{m}_2$ (and thus for any common multiple)

$$A_\mathfrak{m} \cap H_{\mathfrak{m}_1} = A_\mathfrak{m} \cap H_{\mathfrak{m}_2}.$$

It is easy to show that this is an equivalence relation $H_{\mathfrak{m}_1} \sim H_{\mathfrak{m}_2}$ and that the quotient group $A_\mathfrak{m}/H_\mathfrak{m}$ does not depend on the choice of $\mathfrak{m}$. We call the equivalence class of such quotient groups a **congruence divisor class group** A/H, and we call the least modulus $\mathfrak{m}$ for which it can be realized the **conductor** $\mathfrak{f}$ of A/H.

A finite algebraic extension K of k is called a **class field** for A/H if the prime ideals $\mathfrak{p}$ in k which split completely in K (that is, those which have

$e_{K/k} = f_{K/k} = 1$ for each prime $\mathfrak{P}$ of K above $\mathfrak{p}$) are precisely those which belong to H. For the real infinite places in $\mathfrak{f}$ this needs an elucidation: we require that the extensions to K of a real infinite place in k are all complex if it is in $\mathfrak{f}$ and all real otherwise. We can now state the key results of the classical theory.

Theorem 32 *Every abelian extension K/k is a class field for some A/H. Conversely, there corresponds to each congruence divisor class group A/H in k a unique class field K/k, and K is abelian over k.*

In stating the next two theorems we shall denote by K the class field corresponding to A/H in k.

Theorem 33 *The places in k which ramify in the extension K/k are precisely those which divide $\mathfrak{f}$. Moreover, if χ runs through the characters of A/H and $\mathfrak{f}_\chi$ is the conductor of χ, then the finite part of $\prod_\chi \mathfrak{f}_\chi$ is just $\mathrm{norm}_{K/k}\mathfrak{d}_{K/k}$, where $\mathfrak{d}_{K/k}$ is the different of the extension.*

To define the **conductor** of χ we form $H^\sharp$, the union of those cosets of H on which $\chi = 1$; the conductor of χ is just the conductor of $A/H^\sharp$.

Theorem 34 *The Galois group of K/k is isomorphic to A/H. This isomorphism is canonically realized by means of the Artin symbol $\left(\frac{K/k}{\mathfrak{a}}\right)$; the map $\mathfrak{a} \mapsto \left(\frac{K/k}{\mathfrak{a}}\right)$ is an epimorphism $A \to \mathrm{Gal}(K/k)$ with kernel H.*

The second sentence is the **Artin Reciprocity Law**. It was only discovered a generation later than the rest of classical class field theory was formulated (and some five years after the rest was proved); but as soon as it was discovered it was recognized as the central result of the theory. It contains all previously known reciprocity laws.

Corollary *Let $\mathfrak{p}$ be a prime ideal in k not dividing $\mathfrak{f}$. Then $\mathfrak{p}$ is a product of g prime factors $\mathfrak{P}$ in K, each with $\mathrm{norm}_{K/k}\mathfrak{P} = \mathfrak{p}^f$ where $fg = (K/k)$ and f is the order in A/H of the ideal class containing $\mathfrak{p}$.*

These theorems enable one to describe the factorization in K of primes in k in terms of objects in k. To do this, we need to find the H associated with a given abelian extension K/k. Since we know the conductor of H (or, more easily, the primes which divide it) by Theorem 33, there are only a small number of possibilities for H; and all but one of these can be eliminated by considering the factorization of a few small unramified primes. (Compare Exercise 5.1.) Implicit in Theorem 34 is the generalization of Dirichlet's theorem on primes in arithmetic progression; the deduction of this from Theorem 34 can be found in §16.

Theorem 35 *Each coset of H in A contains an infinity of prime ideals.*

As the theory developed, it became clear that Hilbert's original question was somewhat peripheral. But the theory does provide an answer to it, as follows.

Theorem 36 *Let H be the group of principal ideals in A. Then $\mathfrak{f} = (1)$ and the class field for A/H is the absolute class field in the sense of Hilbert.*

These theorems do not however tell one anything about the group of units in K or the class number of K; and broadly speaking it is only when one can construct class fields explicitly that one can obtain any information about these. For arbitrary k, the construction of K from H is the one major unsolved problem in the theory; but it should be remembered that in mathematics not all problems have solutions. The simplest examples of what is known are as follows.

According to the Kronecker-Weber Theorem, the abelian extensions of $\mathbf{Q}$ are just the cyclotomic fields — that is, the subfields of some $\mathbf{Q}(\epsilon)$, where ϵ is a root of unity. Analytically $\epsilon = \exp(2\pi i m/n)$ for some m, n in $\mathbf{Z}$, so that ϵ can be regarded as the value of the periodic function $\exp(z)$ at a division point — that is, at a value of z which is a rational submultiple of its period $2\pi i$. In the 1890s elliptic modular functions were a major growth industry (as they have now become again); and the work of Kronecker and Weber, and of their students, strongly suggested that all abelian extensions K of complex quadratic fields k could be generated by values of certain of these functions at distinguished points, and that if $\mathfrak{p}$ is a prime ideal in a complex quadratic field k then the factorization of $\mathfrak{p}$ in K depends only on congruence properties of $\mathfrak{p}$ in k; but a complete proof of these assertions had to await the work of Takagi. These results have been generalized to other (but still rather limited) types of algebraic number field k by Shimura and others, using the theory of complex multiplication of Abelian varieties

Some of the results above can be put into analytic form, in terms of L-series and zeta functions. (There was a period when strenuous attempts were made to rid class field theory of all taint of analysis, but the pendulum has now swung back and the importance of the analytic aspect is recognized.) Let s be a complex variable, and take $\Re s > 1$ so that all problems of convergence are trivial. Let χ be any character on A/H with conductor f_χ. We can lift χ back to a congruence character $\mod f_\chi$, that is, a homomorphism

$$\chi : A_f \to \{\text{complex numbers of absolute value 1}\}$$

whose kernel contains H_f^0. Set $\chi(\mathfrak{a}) = 0$ for all integral ideals $\mathfrak{a}$ not in A_f and write

$$L(s, \chi) = \sum \frac{\chi(\mathfrak{a})}{(\text{Norm } \mathfrak{a})^s} = \prod \left(1 - \frac{\chi(\mathfrak{p})}{(\text{Norm } \mathfrak{p})^s}\right)^{-1}$$

where the sum is over all integral ideals $\mathfrak{a}$ in k and the product is over all prime ideals $\mathfrak{p}$ in k. This is the classical (Dirichlet) L-series; the equality of sum and product is just the Unique Factorization Theorem. In the particular case when χ is the principal character, so that $f_\chi = (1)$, we obtain the Dedekind zeta function

$$\zeta_k(s) = \sum (\text{Norm } \mathfrak{a})^{-s} = \prod (1 - (\text{Norm } \mathfrak{p})^{-s})^{-1}$$

where the sum is taken over all integral non-zero ideals $\mathfrak{a}$ and the product over all prime ideals $\mathfrak{p}$. In this language we can restate Theorem 33 and the Corollary to Theorem 34 together in the following form.

Theorem 37 $\zeta_K(s) = \prod L(s, \chi)$ *where the product is over all characters of A/H.*

Theorems 32 to 34 are the main theorems of classical class field theory; it was essentially a global theory, since the importance of local class field theory did not become apparent till later. Closely linked to the development of classical class field theory was the study, largely inspired by Hasse, of local-to-global questions. These can appear in two ways, though they are not essentially different:

(i) If a condition is satisfied (for example, if an equation is soluble) in every k_v, does that imply that it is satisfied in k?

(ii) Suppose we have a set of objects O defined with respect to k, such that the completion map $k \to k_v$ sends O to an object O_v defined with respect to k_v. Given a collection of O_v for v in S, where S is some subset of the set of all valuations, is there an O of which they are all images?

In each case, if the answer is negative, can we describe the obstruction? The simplest example of a positive theorem which answers a question of this kind is the Hasse Norm Theorem:

Theorem 38 *Suppose that K/k is cyclic and α in k is everywhere locally a norm for K/k — that is, α is in the image of the map $K \otimes_k k_v \to k_v$ induced for each v by completion from the norm map $K \to k$ — then α is the norm of an element of K.*

If K/k is merely abelian then this result need not be true; for example, if we take $k = \mathbf{Q}$, $K = \mathbf{Q}(\sqrt{13}, \sqrt{17})$ then -1 is a norm everywhere locally but it is not a global norm.

18 Chevalley's reformulation

It still remains true that for many applications the classical language is the most convenient. But there were several unsatisfactory features about the original formulation of class field theory. That it only gives information about abelian extensions is probably a fact of life; though there is now a subject called non-abelian class field theory, it does not provide answers to the most obvious questions — perhaps because there are no good answers to them. But sometimes (as with Theorem 38) the classical theory only gave information about a restricted class of abelian extensions. There were also aesthetic objections: to the substantial role which analysis played in the original proofs, to the ugly definition of the congruence divisor class group A/H and to the complications of the group-theoretic arguments involved in the original proofs.

These were the reasons for a reformulation in terms of homological algebra. This produced some new theorems which could be stated in classical language — for example it provided a description of the group

$$\{\text{everywhere local norms}\}/\{\text{actual norms}\}$$

which is the obstruction to the obvious generalisation of Theorem 38. (Indeed, one now instinctively assumes that all obstructions are best described in term of cohomology groups.) In this language the classical theorems became statements about certain cohomology groups, which remained valid for normal extensions which were not necessarily abelian; however, it was only in the abelian case that these cohomology groups could be described in down-to-earth terms. At the same time Chevalley introduced the concept of idèles. In the language of idèles A/H can be replaced as follows. Let $\widehat{H}$ be any open subgroup of J_k containing k^*; by definition $\widehat{H}$ has finite index in J_k. Chevalley showed that there is a natural one-one correspondence between pairs A, H and subgroups $\widehat{H}$, and that this enables one to reformulate most of the main theorems of classical class field theory.

This led to a more radical reformulation, which also takes into account infinite abelian extensions. Let $\mathcal{S}$ be a finite set of valuations which contains all Archimedean valuations; $\mathcal{S}$ is related to, but is not necessarily the same as, the set of places which occur in $\mathfrak{m}$ in Weber's definitions. Denote by $I^{\mathcal{S}}$ the group of those fractional ideals which do not involve any prime ideal in $\mathcal{S}$ in their factorization, and by $J^{\mathcal{S}}$ the group of those idèles whose components are 1 at each place in $\mathcal{S}$. These are subgroups of I_k, J_k respectively, and $J^{\mathcal{S}}$ is closed in J_k. The natural map $J_k \to I_k$ already introduced on page 50 induces a continuous map $J^{\mathcal{S}} \to I^{\mathcal{S}}$, which we write in the form $\xi \mapsto (\xi)^{\mathcal{S}}$. It is also convenient to write $F_{K/k}$ for the Artin map $\mathfrak{a} \mapsto \left(\frac{K/k}{\mathfrak{a}}\right)$, which is a homomorphism $I^{\mathcal{S}} \to \mathrm{Gal}(K/k)$ provided $\mathcal{S}$ contains all ramified primes. Theorem 34 can now be rewritten in the following apparently weaker form, in which we fix an abelian extension K/k and $\mathcal{S}$ is assumed to contain all the primes which ramify in K/k.

Theorem 39 *There exists $\epsilon > 0$ such that $F_{K/k}((\xi))$ is the identity for all ξ in k^* for which $\|\xi - 1\|_v < \epsilon$ for each v in $\mathcal{S}$.*

For suitable ϵ these conditions imply $\xi \equiv 1 \bmod \mathfrak{m}$, so that Theorem 39 does follow from Theorem 34. The converse, that Theorem 39 implies Theorem 34, is less easy. The key fact is that $F_{K/k}$ is trivial for all ξ which are norms for K/k, and hence (after Theorem 39) for all ξ which are locally norms for every v in $\mathcal{S}$. Now Theorem 34 follows from local class field theory, which is much easier than the global theory.

At this stage Hecke's Grössencharakters enter the picture. These were originally introduced as the most general homomorphisms

$$\chi : I^{\mathcal{S}} \to \{\text{complex numbers of absolute value 1}\}$$

for which Hecke's proof of the functional equation for the associated L-series is valid. One form of the condition on χ is (91), as was shown in Chapter 4; this is equivalent to the apparently weaker condition that for any neighbourhood $\mathcal{N}$ of 1 there exists an $\epsilon > 0$ such that the inverse image of $\mathcal{N}$ contains all principal ideals (ξ) with ξ satisfying $\|\xi - 1\|_v < \epsilon$ for each v in $\mathcal{S}$. (See Exercise 4.3.) This led to the idea of an admissible map. Let G be any commutative topological group; then a homomorphism $\phi : I^{\mathcal{S}} \to G$ is called **admissible** if to any neighbourhood $\mathcal{N}$ of the identity in G there corresponds an $\epsilon > 0$ such that $\phi((\xi))$ is in $\mathcal{N}$ for any ξ in k with $\|\xi - 1\|_v < \epsilon$ for each v in $\mathcal{S}$. Theorem 39 can now be rewritten in the form

Theorem 40 *The Artin map $F_{K/k} : I^S \to \mathrm{Gal}(K/k)$ is admissible.*

So far, we have been considering a single abelian extension K/k. If instead we consider all finite extensions of k and the corresponding Artin maps, it becomes evident that the natural language to use is the language of idèles. The detailed translation depends on

Lemma 35 *Suppose that G is topologically complete, and that the homomorphism $\phi : I^S \to G$ is admissible. Then there is a unique continuous homomorphism $\psi : J_k \to G$ such that*

(i) *$\psi(\xi) = 1$ for each ξ in k^*,*

(ii) *$\psi(\boldsymbol{\xi}) = \phi((\boldsymbol{\xi})^S)$ for each $\boldsymbol{\xi}$ in J^S.*

Conversely, let ψ be any continuous homomorphism $J_k \to G$ which is trivial on k^*, and suppose that there is a neighbourhood of the identity in G which contains no non-trivial subgroup; then ψ comes from some set S and some associated admissible map $\phi : I^S \to G$.

Such a ψ induces a continuous homomorphism $C_k = J_k/k^* \to G$ where C_k is called the **idèle class group**; we call this map ψ too. In particular, Theorem 40 is equivalent to the existence of a ψ which also satisfies

$$\psi(\boldsymbol{\xi}) = F_{K/k}((\boldsymbol{\xi})^S) \quad \text{for all} \quad \boldsymbol{\xi} \text{ in } J^S;$$

such a ψ, which is unique by the lemma, is called the Artin map $\psi_{K/k}$.

Now the main theorems of class field theory take the following form.

Theorem 41 *To every finite abelian extension K/k corresponds an Artin map $\psi : J_k \to \mathrm{Gal}(K/k)$. This is an epimorphism whose kernel is precisely the open subgroup $k^*(\mathrm{Norm}_{K/k}(J_K))$; and it induces an isomorphism*

$$C_k/\mathrm{Norm}_{K/k}(C_K) \approx \mathrm{Gal}(K/k).$$

If $L \supset K \supset k$ with L abelian over k, then there is a commutative diagram

$$
\begin{array}{ccc}
C_k & \xrightarrow{\psi_{L/k}} & \mathrm{Gal}(L/k) \\
\| & & \downarrow \\
C_k & \xrightarrow{\psi_{K/k}} & \mathrm{Gal}(K/k)
\end{array}
$$

where the right hand arrow is restriction to K. Conversely, if $\mathcal{N} \supset k^*$ is an open subgroup of J_k then there is a unique abelian extension K/k such that $\mathcal{N}$ is $k^*(\mathrm{Norm}_{K/k}(J_K))$, the kernel of the Artin map $\psi_{K/k}$.

The commutative diagram above enables us to take the inverse limit as K runs through all abelian extensions of k. In this way we obtain a continuous homomorphism

$$\psi_k : \mathcal{C}_k \to \mathrm{Gal}(k^{\mathrm{ab}}/k)$$

where k^{ab} is the maximal abelian extension of k and the Galois group has the topology proper to an inverse limit — that is to say, a base for the open neighbourhoods of the identity is given by the subgroups of finite index. The map ψ_k is onto, and its kernel is just the connected component of $\mathcal{C}_k$. Unfortunately, for general k very little is known about this kernel.

19 Reciprocity theorems

It is natural to ask whether the Quadratic Reciprocity Law (Theorem 25) can be generalized to arbitrary algebraic number fields k and to higher powers. This is possible, subject to one important restriction on k. Throughout this section, $m > 1$ will be a fixed integer and we shall require k to contain the m-th roots of unity. Denote by S the set consisting of the infinite places of k and those primes of k which divide m; and note that if $\mathfrak{p}$ is not in S then $X^m - 1$ splits completely in k and therefore in $\mathfrak{o}_\mathfrak{p}/\mathfrak{p}$, whence $\mathrm{Norm}\,\mathfrak{p} \equiv 1 \bmod m$.

For α in k^* let $K = k(\epsilon)$ where $\epsilon^m = \alpha$; denote by $S(\alpha)$ the union of S and the set of primes at which α is not a unit, and define the **power residue symbol** $\left(\frac{\alpha}{\mathfrak{p}}\right)$ by

$$\left(\frac{K/k}{\mathfrak{p}}\right)\epsilon = \left(\frac{\alpha}{\mathfrak{p}}\right)\epsilon$$

for any prime $\mathfrak{p}$ of k which is not in $S(\alpha)$. If we extend this definition by multiplicativity to $\left(\frac{\alpha}{\mathfrak{b}}\right)$ for all $\mathfrak{b}$ in $I^{S(\alpha)}$ then $\left(\frac{\alpha}{\mathfrak{b}}\right)$ is an m-th root of unity which is unaltered by replacing ϵ by another m-th root of α. We call $\left(\frac{\alpha}{\mathfrak{p}}\right)$ a power residue symbol because $\left(\frac{\alpha}{\mathfrak{p}}\right) = 1$ is equivalent to $\epsilon^{\mathrm{Norm}\,\mathfrak{p}} \equiv \epsilon \bmod \mathfrak{p}$, hence to $\epsilon \bmod \mathfrak{p}$ lying in $\mathfrak{o}_\mathfrak{p}/\mathfrak{p}$ and so by Hensel's Lemma to α being an m-th power in $k_\mathfrak{p}$.

Now suppose that α' is also in k and that $K' = k(\epsilon')$ with $(\epsilon')^m = \alpha'$, and write $L = KK'$. By Theorem 16 $\left(\frac{K/k}{\mathfrak{p}}\right)$ is the restriction of $\left(\frac{L/k}{\mathfrak{p}}\right)$ to K and so on; and therefore

$$\left(\frac{\alpha\alpha'}{\mathfrak{p}}\right)\epsilon\epsilon' = \left(\frac{L_0/k}{\mathfrak{p}}\right)\epsilon\epsilon' = \left(\frac{L/k}{\mathfrak{p}}\right)\epsilon\epsilon'$$
$$= \left\{\left(\frac{K/k}{\mathfrak{p}}\right)\epsilon\right\}\left\{\left(\frac{K'/k}{\mathfrak{p}}\right)\epsilon'\right\} = \left(\frac{\alpha}{\mathfrak{p}}\right)\left(\frac{\alpha'}{\mathfrak{p}}\right)\epsilon\epsilon'$$

where $L_0 = k(\epsilon\epsilon')$. It follows that

$$\left(\frac{\alpha\alpha'}{\mathfrak{p}}\right) = \left(\frac{\alpha}{\mathfrak{p}}\right)\left(\frac{\alpha'}{\mathfrak{p}}\right) \tag{101}$$

provided the right hand side is defined. Again $(\frac{K/k}{\mathfrak{p}})\epsilon \equiv \epsilon^{\mathrm{Norm}\,\mathfrak{p}} \bmod \mathfrak{p}$ implies

$$\left(\frac{\alpha}{\mathfrak{p}}\right) \equiv \alpha^{(\mathrm{Norm}\,\mathfrak{p}-1)/m} \bmod \mathfrak{p}. \tag{102}$$

This determines $(\frac{\alpha}{\mathfrak{p}})$ uniquely as an m-th root of unity; moreover it gives an alternative proof of (101), and it shows that if $\mathfrak{b}$ is integral then $(\frac{\alpha}{\mathfrak{b}})$ only depends on $\alpha \bmod \mathfrak{b}$.

Now let A/H be the congruence divisor class group corresponding to K/k. It follows from Theorem 33 that the only finite primes which can appear in $\mathfrak{f}$, the conductor of A/H, are those in $S(\alpha)$. Let γ be in k^*; then γ is in H if γ is in $(k_v^*)^m$ for each v in $S(\alpha)$, and it follows from Theorem 34 that

$$\left(\frac{\alpha}{\gamma\mathfrak{b}}\right) = \left(\frac{\alpha}{\mathfrak{b}}\right).$$

The theory can be given a more symmetric form in terms of the **Hilbert norm residue symbol**. This was introduced by Hilbert in the case $m = 2$, by means of the much simpler definition (105) below, and by Hasse in general. The first step is to define the local Artin map corresponding to the abelian extension $k(\epsilon)/k$; because it costs nothing, we shall do this for an arbitrary abelian extension K/k. Let $G = A/H$ be a congruence divisor class group in k, with conductor $\mathfrak{f}$, and let S be the set of prime ideals which divide $\mathfrak{f}$. Denote by $I^S = A$ the group of those fractional ideals whose prime factorizations involve no ideal in S, and by J^S the group of idèles whose components are 1 at each place in S.

Lemma 36 *In the notation above, for each place v of k there is a unique homomorphism $f_v : k_v^* \to G$ such that*

(i) *if $\mathfrak{p}$ is a prime ideal not in S then $f_{\mathfrak{p}}$ is trivial on $\mathfrak{o}_{\mathfrak{p}}^*$,*

(ii) *$f = \prod f_v$ is well-defined and continuous on J_k,*

(iii) *if ξ is in J^S then $f((\xi))$ is the image of (ξ) in G, where (ξ) is as on page 50,*

(iv) *$f(\beta) = 1$ for each β in k^*.*

Proof If $\mathfrak{p}$ is not in S we must take $f_{\mathfrak{p}}(\alpha_{\mathfrak{p}})$ to be the image of $\mathfrak{p}^n$ in G where $\mathfrak{p}^n \| \alpha_{\mathfrak{p}}$; this ensures (i) and (iii), and also (ii) provided the f_v for v in S

are continuous. Let $N > 0$ annihilate G and temporarily let the f_v with v in S be any homomorphisms $k_v^* \to G$. Then $f(\beta)$ is the identity provided that β is in k^* and in $(k_v^*)^N$ for each v in S, for then Theorem 17 implies that we can write $\beta = \beta_1^N \beta_2$ with β_1 in J_k, β_2 in k^* and $\beta_2 \equiv 1 \bmod \mathfrak{f}$ — so that (β_2) is in H. The natural map $k^* \to \prod_{v \text{ in } S}(k_v^*/(k_v^*)^N)$ is onto, using Theorem 17 again, and it follows that there is just one choice of the f_v with v in S for which (iv) holds. Moreover these f_v are continuous because the $k_v^*/(k_v^*)^N$ are finite.   □

The local Artin map $\psi_v : k^* \to A/H \to \mathrm{Gal}(K/k)$ is obtained by composing f_v with the isomorphism of Theorem 34. It also follows from the corollary to that theorem that if α is a norm for K/k and v is not is S then $f_v(\alpha)$ is the identity. Using Theorem 17 yet again, the construction in the proof of the lemma implies that this holds also if v is in S.

The **Hilbert symbol** $(\alpha, \beta)_v$ for α, β in k^* is now defined by

$$\psi_v(\beta)\epsilon = (\alpha, \beta)_v\,\epsilon;$$

we shall shortly prove that it extends to a continuous function on k_v^*. The value of $(\alpha, \beta)_v$ is again an m-th root of unity, and is unaltered by replacing ϵ by another m-th root of α. It is trivial that

$$(\alpha, \beta\beta')_v = (\alpha, \beta)_v\,(\alpha, \beta')_v$$

and an argument like that which was used to prove (101) gives

$$(\alpha\alpha', \beta)_v = (\alpha, \beta)_v\,(\alpha', \beta)_v.$$

Lemma 37 *Each of the symbols $(\alpha, -\alpha)_v$ and $(\alpha, 1 - \alpha)_v$ is equal to 1 whenever it is defined.*

Proof Write $d = m/[K : k]$, so that the conjugates of ϵ over k are the $\zeta^\mu \epsilon$ with $d | \mu$ where ζ is a primitive m-th root of unity. For any γ in k,

$$\gamma^m - \alpha = \prod_{\mu=0}^{m-1}(\gamma - \zeta^\mu \epsilon) = \mathrm{norm}_{K/k}\left\{\prod_{\nu=0}^{d-1}(\gamma - \zeta^\nu \epsilon)\right\};$$

so $\psi_v(\gamma^m - \alpha)$ is the identity and $(\alpha, \gamma^m - \alpha)_v = 1$. Now set $\gamma = 0, 1$.   □

Corollary *For any α, β in k^* we have $(\alpha, \beta)_v\,(\beta, \alpha)_v = 1$.*

Proof By bilinearity

$$(\alpha\beta, -\alpha\beta)_v = (\alpha, -\alpha)_v\,(\beta, -\beta)_v\,(\alpha, \beta)_v\,(\beta, \alpha)_v$$

and all but the last two terms are equal to 1.   □

It follows that the Hilbert symbol induces a skew-symmetric bilinear map

$$k_v^*/(k_v^*)^m \times k_v^*/(k_v^*)^m \to \{m - \text{th roots of unity}\}. \tag{103}$$

For by the definition of ψ_v we can take the second argument to be in k_v^*, and even in $k_v^*/(k_v^*)^m$ because the Hilbert symbol is killed by m. But

$$k^*/(k^* \cap (k_v^*)^m) \to k_v^*/(k_v^*)^m$$

is onto, so corresponding to any β in k_v^* we can find β_1 in k^* such that $(\alpha, \beta)_v = (\alpha, \beta_1)_v$ for all α in k^*. Thus we can use the corollary to extend the range of α from k^* to k_v^*.

The other key property of the Hilbert symbol is

$$\prod_v (\alpha, \beta)_v = 1, \tag{104}$$

which follows from (iv) of Lemma 36. The product is well-defined because $(\alpha, \beta)_v = 1$ whenever α, β are both in $\mathfrak{o}_v^*$ and v is not in S.

Theorem 42 *The Hilbert symbol, regarded as a bilinear form* (103), *is non-degenerate.*

This is a key result in local class field theory. For v in S we cannot prove it in this book, essentially because of the rather indirect construction of f_v in the proof of Lemma 36. But if $v = \mathfrak{p}$ is not in S the result is trivial. For suppose α is in $\mathfrak{o}_\mathfrak{p}^*$ and $\mathfrak{p}\|\pi$; then by construction $(\alpha, \pi)_\mathfrak{p} = 1$ if and only if $\mathfrak{p}$ splits completely in $k(\epsilon)/k$, which by Theorem 19 is the same as saying that α is in $(\mathfrak{o}_\mathfrak{p}^*)^m$. If $\mathfrak{p}^n\|\alpha$ with $m|n$ but α is not in $(k_\mathfrak{p}^*)^m$ then $(\alpha, \pi)_\mathfrak{p} \neq 1$; if $m \nmid n$ and r is the highest common factor of m and n, then $(\alpha, \beta)_\mathfrak{p} \neq 1$ if β is in $\mathfrak{o}_\mathfrak{p}^*$ and not an (m/r)-th power.

Theorem 43 *Let L be abelian over k_v; then the group of norms from L to k_v has index $[L : k_v]$ in k_v^*.*

This result again belongs to local class field theory, though elementary proofs exist. We quote it because the special case $[L : k_v] = 2$ implies the important formula (105) below; but even in this case the elementary proof is no more than a tedious and unilluminating verification.

It is reasonably easy to evaluate the Hilbert symbol when v is not in S. If v is Archimedean the fact that k contains the m-th roots of unity implies that $(\alpha, \beta)_v = 1$ except perhaps when $m = 2$, v is real and α, β are both negative; in this exceptional case (105) shows that $(\alpha, \beta)_v = -1$. The non-Archimedean cases are covered by the following lemma. One relatively simple case for which v is in S is given in (105), and another can be found in the exercise later in this section. For any particular k the Hilbert symbol is given by a table which can easily be constructed with the help of (104). For an illustration of this, see Exercise 5.5.

Lemma 38 *Suppose that $\mathfrak{p}$ is a finite prime of k not in $\mathcal{S}$; then*

$$(\alpha, \beta)_\mathfrak{p} = \left(\tfrac{\gamma}{\mathfrak{p}}\right) \quad \text{where} \quad \gamma = (-1)^{v(\alpha)v(\beta)}\alpha^{v(\beta)}\beta^{-v(\alpha)}.$$

Here γ is a unit at $\mathfrak{p}$, so that the value of $\left(\tfrac{\gamma}{\mathfrak{p}}\right)$ is given by (102).

Proof It follows from the construction of $\psi_\mathfrak{p}$ that $(\alpha, \beta)_\mathfrak{p} = 1$ for α, β in $\mathfrak{o}_\mathfrak{p}^*$; and if $\mathfrak{p}\|\pi$ then $(\pi, -\pi)_\mathfrak{p} = 1$ by Lemma 37, and $(\alpha, \pi)_\mathfrak{p} = \left(\tfrac{\alpha}{\pi}\right)$ by definition if α is in $\mathfrak{o}_\mathfrak{p}^*$. Write $\alpha = \pi^{v(\alpha)}\alpha_0$, $\beta = \pi^{v(\beta)}\beta_0$. Using the results of the first sentence and the bilinearity of the Hilbert symbol

$$(\alpha, \beta)_\mathfrak{p} = (\pi, \beta_0)_\mathfrak{p}^{v(\alpha)}(\alpha_0, \pi)_\mathfrak{p}^{v(\beta)}(-1, \pi)_\mathfrak{p}^{v(\alpha)v(\beta)}$$

and the result now follows from the Corollary to Lemma 37. $\square$

For any α, β in k^* denote by $(\beta)^{\mathcal{S}(\alpha)}$ the ideal obtained from (β) by deleting any prime factors which lie in $\mathcal{S}(\alpha)$, and write

$$\left(\frac{\alpha}{\beta}\right) = \left(\frac{\alpha}{(\beta)^{\mathcal{S}(\alpha)}}\right);$$

note that despite (101) it is not always true that $\left(\tfrac{\alpha}{\beta\beta'}\right) = \left(\tfrac{\alpha}{\beta}\right)\left(\tfrac{\alpha}{\beta'}\right)$. Then we have the general power reciprocity law:

Lemma 39 *If α, β are in k^* then*

$$\left(\frac{\alpha}{\beta}\right)\left(\frac{\beta}{\alpha}\right)^{-1} = \prod(\beta, \alpha)_v$$

where the product is taken over all v in $\mathcal{S}(\alpha) \cap \mathcal{S}(\beta)$.

Proof Using the Corollary to Lemma 37 we have

$$\left(\frac{\alpha}{\beta}\right)\left(\frac{\beta}{\alpha}\right)^{-1} = \left\{\prod_{\mathfrak{p} \text{ not in } \mathcal{S}(\alpha)}(\alpha, \beta)_\mathfrak{p}\right\}\left\{\prod_{\mathfrak{p} \text{ not in } \mathcal{S}(\beta)}(\alpha, \beta)_\mathfrak{p}\right\}$$

$$= \prod_{\mathfrak{p} \text{ not in } \mathcal{S}(\alpha) \cap \mathcal{S}(\beta)}(\alpha, \beta)_\mathfrak{p}$$

because the terms which occur in both products are each equal to 1 as in the proof of Lemma 38. Now use (104). $\square$

Now suppose that $m = 2$, but place no restriction on k. The case $m = 2$ is particularly favourable, because in order to give a formula for $(\alpha, \beta)_v$ it is enough to give a necessary and sufficient condition for $(\alpha, \beta)_v = 1$. Such a condition is as follows:

$$(\alpha, \beta)_v = \begin{cases} 1 & \text{if } \alpha X^2 + \beta Y^2 = 1 \text{ is soluble in } k_v, \\ -1 & \text{otherwise.} \end{cases} \tag{105}$$

This follows easily from Theorems 42 and 43. For we can assume that α is not in $(k_v^*)^2$ and write $L = k_v(\sqrt{\alpha})$. The proof of Lemma 37 shows that $(\alpha, \beta)_v = 1$ if β is a norm for L/k_v, which is equivalent to the solubility of $\alpha X^2 + \beta Y^2 = 1$; and Theorem 43 shows that for fixed α the β for which this holds form a subgroup of index 2 in k_v^*. It now follows from Theorem 42 that $(\alpha, \beta)_v = -1$ if β is not a norm.

What underlies this result is the much more general fact that central simple algebras over k can be classified by means of the Hilbert symbol; see [Weil], Chapters IX to XI or [CF], pages 137-8. Denote by $\mathcal{A}$ the quaternion algebra with norm form

$$X_0^2 - \alpha X_1^2 - \beta X_2^2 + \alpha\beta X_3^2; \tag{106}$$

then each of the two following conditions is necessary and sufficient for $\mathcal{A} \otimes_k k_v$ to be isomorphic to the algebra of 2×2 matrices with elements in k_v:

(i) the quadratic form (106) has a non-trivial zero in k_v;
(ii) $(\alpha, \beta)_v = 1$.

It is easy to see that (i) is equivalent to the solubility of $\alpha X^2 + \beta Y^2 = 1$ in k_v.

Exercise Let $m = p$ be an odd prime and let $k = \mathbf{Q}(\zeta)$ where ζ is a primitive p-th root of unity; thus the only finite place in $\mathcal{S}$ is $\mathfrak{p} = (\pi)$ where $\pi = 1 - \zeta$. The object of this exercise is to provide as Lemma 40 a straightforward way of computing the Hilbert symbol $(\alpha, \beta)_\mathfrak{p}$.

Write $\eta_r = 1 - \pi^r$ for $r = 1, 2, \ldots$, let U_r be the group of elements ϵ in $k_\mathfrak{p}^*$ satisfying $\epsilon \equiv 1 \bmod \pi^r$ and let $U_0 = \mathfrak{o}_\mathfrak{p}^*$. Show that

(i) the image of π generates $k_\mathfrak{p}^*/\mathfrak{o}_\mathfrak{p}^*$,
(ii) $\mathfrak{o}_\mathfrak{p}^* = (\mathfrak{o}_\mathfrak{p}^*)^p U_1$,
(iii) for each r the image of η_r generates U_r/U_{r+1},
(iv) every element of U_{p+1} is in U_1^p. [Use Lemma 20.]

Deduce that π and $\eta_1, \ldots, \eta_p$ generate $k_\mathfrak{p}^*/(k_\mathfrak{p}^*)^p$. This means that we need only evaluate $(\alpha, \beta)_\mathfrak{p}$ when each of α, β is π or one of $\eta_1, \ldots, \eta_p$; and this can be done by induction from the following lemma.

Lemma 40 *If* $\mu \geqslant 1$, $\nu \geqslant 1$ *then*

$$(\eta_\mu, \eta_\nu)_{\mathfrak{p}} = (\eta_\mu, \eta_{\mu+\nu})_{\mathfrak{p}} \, (\eta_{\mu+\nu}, \eta_\nu)_{\mathfrak{p}} \, (\pi, \eta_{\mu+\nu})_{\mathfrak{p}}^\nu \, (\eta_\nu, \pi)_{\mathfrak{p}}^\nu \qquad (107)$$

and in particular $(\eta_\mu, \eta_\nu)_{\mathfrak{p}} = 1$ *if* $\mu + \nu > p$. *Moreover* $(\pi, \pi)_{\mathfrak{p}} = 1$ *and*

$$(\eta_\mu, \pi)_\pi = \left\{ \begin{array}{ll} 1 & \text{if } 1 \leqslant \mu < p, \\ \zeta & \text{if } \mu = p. \end{array} \right.$$

Because p is odd, $(\alpha, \alpha)_{\mathfrak{p}} = 1$ for all α in $k_{\mathfrak{p}}^*$ by Lemma 37. For (107) apply Lemma 37 to $\beta = \eta_\nu/\eta_{\mu+\nu}$, noting that $1 - \beta = \pi^\nu \eta_\nu/\eta_{\mu+\nu}$, and use bilinearity. If $\mu + \nu > p$ then $\eta_{\mu+\nu}$ is in $(k_{\mathfrak{p}}^*)^p$ and so all symbols involving it are equal to 1. If $1 \leqslant \mu < p$ it follows from Lemma 37 that $(\eta_\mu, \pi)_{\mathfrak{p}} = (\eta_\mu, \pi^\mu)_{\mathfrak{p}} = 1$; so the only difficulty comes with the very last assertion. Taking $\nu = 1$, $\mu = p - 1$ in (107) gives $(\zeta, \eta_{p-1})_{\mathfrak{p}} = (\pi, \eta_p)_{\mathfrak{p}}$ because each of the other factors is 1. If $\alpha = \zeta$ in (102) the congruence must actually be an equality because it cannot be satisfied otherwise; now use (104) and Lemma 38 to show that

$$(\zeta, \beta)_{\mathfrak{p}} = \prod_{\mathfrak{q}} \zeta^{v(\beta)(\mathrm{Norm}\,\mathfrak{q}-1)/p}$$

for any β in $\mathfrak{o}_{\mathfrak{p}}^*$, where $\mathfrak{q}^{v(\beta)} \| \beta$ and the product is taken over all prime ideals $\mathfrak{q}$ of k other than $\mathfrak{p}$. Finally, prove that under the same conditions

$$\sum_{\mathfrak{q}} v(\beta)(\mathrm{Norm}\,\mathfrak{q} - 1) \equiv \mathrm{Norm}\,(\beta) - 1 \bmod p^2$$

and $\mathrm{Norm}\,(\eta_{p-1}) = \mathrm{Norm}\,((1 - \pi^{p-1})) \equiv 1 - \mathrm{Tr}\,\pi^{p-1} \equiv 1 - p \bmod p^2$. $\qquad \square$

20 The Kronecker-Weber Theorem

It follows from the Corollary to Theorem 27 that every cyclotomic field is a class field; conversely, Theorem 16 and Theorem 27 enable us to compute the Artin symbol for any cyclotomic field, and it follows that to each congruence divisor class group there corresponds a unique cyclotomic field which is a class field for it. In particular, the field of m-th roots of unity corresponds to $H_{m\infty}$. But this is not enough to prove that these are the only class fields over $\mathbf{Q}$. This assertion is the Kronecker-Weber Theorem (Theorem 44). Its proof is elementary but complicated; it exploits many of the ideas in Chapters 1 and 2, as well as Theorem 27 and its Corollary. The fact that every extension of $\mathbf{Q}$ is ramified at some prime (Corollary to Theorem 21) plays a crucial role.

In this section we shall denote by $C(n)$ any cyclic group of order n. Let K be a cyclic extension of $\mathbf{Q}$ with $G = \mathrm{Gal}(K/\mathbf{Q}) \approx C(p^r)$ where p is prime, let G_1 be the unique subgroup of G of index p and k the fixed field of G_1, so that k is cyclic over $\mathbf{Q}$ with $\mathrm{Gal}(k/\mathbf{Q}) = G/G_1 \approx C(p)$. Let ℓ (which may be equal to p) be a prime which ramifies in k. The inertia group of ℓ for k is the entire Galois group G/G_1; since any proper subgroup of G is in G_1, it follows from Theorem 16 that the inertia group of ℓ for $K/\mathbf{Q}$ is the entire Galois group G. Hence ℓ is totally ramified in K — that is, $e = p^r$ and $f = g = 1$. Such primes are the key to the following arguments.

Lemma 41 *Let $K/\mathbf{Q}$ be cyclic with $G = \mathrm{Gal}(K/\mathbf{Q}) \approx C(p^r)$. Then there are a cyclotomic field L and a field K' with $\mathrm{Gal}(K'/\mathbf{Q}) \approx C(p^s)$ and $s \leqslant r$ such that $LK = LK'$ and the only prime totally ramified in $K'/\mathbf{Q}$ is p itself.*

Proof Suppose that $\ell \neq p$ is totally ramified in K; if there is no such ℓ we can take $K' = K$. In the notation of §5, V is trivial because $[K : \mathbf{Q}]$ is prime to ℓ; hence $T = G$ is cyclic of order dividing $\ell - 1$ by Theorem 15, so that $\ell \equiv 1 \bmod p^r$. Now let L be $\mathbf{Q}(\sqrt[\ell]{1})$; then KL is abelian over $\mathbf{Q}$ and its Galois group is a subgroup of $C(p^r) \times C(\ell - 1)$. Let K_1 be the inertia field of ℓ for $KL/\mathbf{Q}$. As before, the corresponding ramification group is trivial, so $\mathrm{Gal}(KL/K_1)$ is cyclic; its order must be a multiple of $\ell - 1$ by Theorem 27 applied to $KL \supset L \supset \mathbf{Q}$. So $\mathrm{Gal}(KL/K_1) \approx C(\ell - 1)$ since $\mathrm{Gal}(KL/\mathbf{Q})$ contains no larger cyclic subgroup than $C(\ell - 1)$. A prime other than ℓ which ramifies in K_1 must ramify in KL and therefore in K, for only ℓ ramifies in L. No prime ramifies in $K_1 \cap L$; so $K_1 \cap L = \mathbf{Q}$, and now $K_1 L = KL$ by a comparison of degrees. But $\mathrm{Gal}(KL/\mathbf{Q})$ has $\mathrm{Gal}(L/\mathbf{Q}) \approx C(\ell - 1)$ as a quotient, so it has the form $C(p^s) \times C(\ell - 1)$ for some $s \leqslant r$; and hence $\mathrm{Gal}(K_1/\mathbf{Q}) \approx C(p^s)$ by Lemma 25.

If K_1 does not have the properties required of K', then we can repeat the process again on K_1. This process must eventually end, because one fewer prime is ramified (not necessarily totally) in K_1 than in K.   □

Lemma 42 (i) *Let p be an odd prime and let K be a field cyclic over $\mathbf{Q}$ with $\mathrm{Gal}(K/\mathbf{Q}) \approx C(p)$. If p is the only prime which ramifies in K then K is a subfield of the field of p^2-th roots of unity.*

(ii) *If $[K : \mathbf{Q}] = 2$ and 2 is the only prime which ramifies in K, then K is $\mathbf{Q}(\sqrt{2})$, $\mathbf{Q}(\sqrt{-1})$ or $\mathbf{Q}(\sqrt{-2})$, all of which are subfields of $\mathbf{Q}(\sqrt[8]{1})$.*

Proof After the results in §10, (ii) is trivial; so we need only consider (i). Let $L = \mathbf{Q}(\sqrt[p]{1})$; then $\mathrm{Gal}(KL/L) \approx C(p)$ since K and L have coprime degrees over $\mathbf{Q}$. By a standard result in Galois theory, $KL = L(\alpha)$ where $\alpha^p = \beta$ is in L. Choose a generator σ of $\mathrm{Gal}(KL/\mathbf{Q}) \approx C(p(p-1))$; then K is the fixed field of σ^p and L is the fixed field of σ^{p-1}. Thus $\sigma^{p-1}\beta = \beta$ and hence $\sigma^{p-1}\alpha = \zeta\alpha$ where $\zeta^p = 1$; but $\zeta \neq 1$ because we can assume that α is not in L. As in §13, we write $\pi = 1 - \zeta$. Since σ induces a generator of $\mathrm{Gal}(L/\mathbf{Q})$ there is a primitive root $n \bmod p$ such that $\sigma\zeta = \zeta^n$ and hence $\sigma^{p-1}(\sigma\alpha) = \sigma(\sigma^{p-1}\alpha) = \zeta^n(\sigma\alpha)$; in other words, $(\sigma\alpha)/\alpha^n$ is invariant under σ^{p-1} and is therefore in L, and hence $(\sigma\beta)/\beta^n$ is a p-th power in L.

It follows that $(\sigma\alpha)/\alpha$ is not in L and therefore generates KL over L. The prime ideal (π) in L is invariant under σ, so $(\sigma\beta)/\beta$ is a π-adic unit. Replacing α by $(\sigma\alpha)/\alpha$ and β by $(\sigma\beta)/\beta$, we can assume that β is a π-adic unit, and by further replacing α by $\alpha^{1-p} = \alpha/\beta$ we can also assume that $\beta \equiv 1 \bmod \pi$; this means that $\beta = \zeta^a\gamma$ where $\gamma \equiv 1 \bmod \pi^2$ and therefore

$$\gamma \equiv 1 + c\pi^m \bmod \pi^{m+1}$$

for some $m > 1$ and some c in $\mathbf{Z}$ prime to p. For δ in the local field L_π, Lemma 19 shows that $\delta \equiv 1 \bmod \pi$ is a p-th power in L_π if and only if $\delta \equiv 1 \bmod \pi^p$. Since $(\sigma\gamma)/\gamma^n = (\sigma\beta)/\beta^n$ is a p-th power in L and so also in L_π,

$$\sigma\gamma \equiv \gamma^n \bmod \pi^p.$$

But $\sigma\pi = 1 - \zeta^n \equiv n(1 - \zeta) \bmod \pi^2$ and therefore

$$\sigma\gamma \equiv 1 + cn^m\pi^m \bmod \pi^{m+1}.$$

The last three displayed equations together imply that either $m \geqslant p$ or $cn^m \equiv cn \bmod p$, and the latter also requires $m \geqslant p$ since $m > 1$ and n is a primitive root $\bmod p$. It follows that $\gamma \equiv 1 \bmod \pi^p$.

To prove the lemma, it is enough to show that γ is a p-th power in L. Since $K_1 = L(\sqrt[p]{\gamma})$ is in $K(\sqrt[p^2]{1})$, no prime other than p ramifies in $K_1/\mathbf{Q}$ and $K_1/\mathbf{Q}$ is abelian. But by Lemma 30 and the result above, (π) does not ramify in K_1/L. Let k be the inertia field of p for $K_1/\mathbf{Q}$; then $[K_1 : k] = p - 1$ and $k = \mathbf{Q}$ because no prime ramifies in $k/\mathbf{Q}$. Thus $[K_1 : \mathbf{Q}] = p - 1$ and γ is a p-th power in L.					□

Theorem 44 *Every abelian extension of* $\mathbf{Q}$ *is cyclotomic.*

Proof We prove this by induction on the degree of the extension. Let K be abelian over $\mathbf{Q}$ with $\mathrm{Gal}(K/\mathbf{Q}) = G$. If G is a non-trivial direct product $G_1 \times G_2$ then K is the least field containing the fixed fields of G_1 and of G_2, which are abelian extensions of $\mathbf{Q}$ of degrees lower than that of K. Hence we may assume that $G \approx C(p^r)$ where p is prime. In view of Lemma 41 we can assume that p is the only prime which is totally ramified in K, and in view of Lemma 42 we can further assume that $r > 1$.

Consider first the case when p is odd; let K_1 be the unique subfield of K with $[K_1 : \mathbf{Q}] = p$ and let L be the field of p^{r+1}-th roots of unity. Since $K \cap L \supset K_1$ by Lemma 42 and $\mathrm{Gal}(L/\mathbf{Q}) \approx C(p^r(p-1))$,

$$H = \mathrm{Gal}(KL/\mathbf{Q}) \approx C(p^r(p-1)) \times C(p^{r'}) \text{ for some } r' < r \qquad (108)$$

because H is a proper subgroup of $C(p^r(p-1)) \times C(p^r)$ which admits a $C(p^r(p-1))$ as a quotient group. Let $H_1 = \mathrm{Gal}(KL/L)$; then

$$H/H_1 = \mathrm{Gal}(L/\mathbf{Q}) \approx C(p^r(p-1));$$

by (108) H_1 is a direct factor of H. If $H = H_1 \times H_2$ then the fixed field of H_2 has degree $p^{r'} < p^r$ over $\mathbf{Q}$, and KL is the compositum of it and L. Hence by the induction hypothesis K is cyclotomic.

When $p = 2$ this argument needs some modification. The Galois group over $\mathbf{Q}$ of the field of 2^{r+2}-th roots of unity is $C(2) \times C(2^r)$, and hence this field has a subfield L with $\mathrm{Gal}(L/\mathbf{Q}) \approx C(2^r)$. Now the argument used for p odd will still work, provided we show that the unique quadratic subfield of K is $\mathbf{Q}(\sqrt{2})$; for since L satisfies all the conditions imposed on K, this will hold for L also. If K is real this is trivial. If K is complex it admits complex conjugacy as a non-trivial automorphism of order 2; all its proper subfields are fixed under this and hence are totally real. $\qquad \square$

Appendix

A1 Prerequisites

In this section we prove the standard results about finitely generated abelian groups and lattices, and about norms and traces in field extensions of finite degree, for the benefit of those readers who do not know them already. They are prerequisites for everything in this book. We also give a brief account, without proofs, of the definition and key properties of Haar measure. Logically, this is not essential for the arguments in most of the book, but it underpins the point of view adopted. For a really single-minded reliance on Haar measure, see [Weil].

A1.1 Finitely generated abelian groups and lattices

In this subsection, all abelian groups will be written additively.

Lemma A1 *Let G be a finitely generated torsion-free abelian group such that*

(i) *G is generated by $x_1, \ldots, x_n$,*
(ii) *G cannot be generated by less than n elements.*

Then there is no non-trivial relation $a_1 x_1 + \cdots + a_n x_n = 0$ with the a_ν in **Z**.

Proof Assume that there is such a relation, and among all sets of n generators and all non-trivial relations as above choose that one for which $A = |a_1| + \cdots + |a_n|$ is least. We have two cases, either of which will give a contradiction.

(a) At least two of the a_ν are non-zero. By permuting subscripts and changing signs we can assume that $a_1 \geqslant a_2 > 0$. Now consider the

117

base $x_1, x_1 + x_2, x_3, \ldots$ for G. In terms of this base our relation becomes $(a_1 - a_2)x_1 + a_2(x_1 + x_2) + \cdots = 0$, and this has a smaller value of A.

(b) Only one a_ν is non-zero. We can take this to be a_1, so that the relation is $a_1 x_1 = 0$. Since G is torsion-free, $x_1 = 0$; so G is generated by the $n - 1$ elements $x_2, \ldots, x_n$.

These contradictions prove the lemma. $\qquad\qquad\qquad\qquad\qquad\qquad \square$

We call a base having the properties in Lemma A1 a **minimal base**. Every finitely generated torsion-free abelian group has a minimal base, for we need only choose a base consisting of as few elements as possible.

Lemma A2 *Let G be a finitely generated torsion-free abelian group, and let H be a subgroup of G. Then there exist a minimal base $x_1, \ldots, x_n$ of G and integers $m_1, \ldots, m_r$ for some $r \leqslant n$ such that*

(i) *the m_ρ are positive and $m_\rho | m_{\rho+1}$ for $\rho = 1, \ldots, r - 1$,*

(ii) *$m_1 x_1, \ldots, m_r x_r$ are a minimal base for H.*

If H has finite index in G then $r = n$.

Proof The proof is really by induction on n, but because of the form of the induction hypothesis it is better not to state it in that way. We can assume that H is non-trivial. Let $y_1, \ldots, y_n$ be any minimal base for G and let $h = a_1 y_1 + \cdots + a_n y_n$ be any non-zero element of H. Making an integral unimodular transformation on the y_ν cannot decrease the highest common factor of the a_ν and is an action which can be reversed; so it leaves the highest common factor unchanged. Now choose h to be a non-zero element of H for which this highest common factor is as small as possible, and then choose that minimal base $y_1, \ldots, y_n$ for G for which $A = |a_1| + \cdots + |a_n|$ is as small as possible for this particular h. If two or more of the a_ν were non-zero, we could decrease A just as in the proof of Lemma A1; hence after permuting subscripts we can assume that $a_1 = m_1 > 0$ and all other a_ν vanish. Let $z = b_1 y_1 + \cdots + b_n y_n$ be any element of H; then

- $m_1 | b_1$ since if $0 < b_1 - cm_1 < m_1$ for some c in $\mathbf{Z}$ then $z - ch$ would give rise to a smaller highest common factor than h,

- $b_2 y_2 + \cdots + b_n y_n = z - (b_1/m_1)h$ is in H,

- $m_1 | b_\nu$ for each ν, for otherwise $m_1 y_1 + b_2 y_2 + \cdots + b_n y_n$ would give rise to a smaller highest common factor than h.

Now let G_1 be generated by $y_2, \ldots, y_n$, let $H_1 = G_1 \cap H$ and write $x_1 = y_1$. We have shown that

$$G = \{x_1\} \oplus G_1, \quad H = \{m_1 x_1\} \oplus H_1.$$

If we repeat the process on G_1 and H_1, we shall have $m_1 | m_2$ by the last bullet point. We stop when H is exhausted. If $r < n$ then all the $m x_n$ would generate distinct cosets of H, and H would not have finite index in G. □

Theorem A1 *Let G be a finitely generated abelian group. There exist elements $x_1, \ldots, x_n$ of G and integers $m_1, \ldots, m_r$ for some $r \leqslant n$ such that*

(i) $m_\rho x_\rho = 0$ *for $\rho = 1, \ldots, r$,*

(ii) *each $m_\rho > 1$ and $m_\rho | m_{\rho+1}$ for $\rho = 1, \ldots, r - 1$,*

(iii) *each element of G can be written uniquely as $a_1 x_1 + \cdots + a_n x_n$ where the a_ν are integers and $0 \leqslant a_\rho < m_\rho$ for $\rho = 1, \ldots, r$.*

Moreover G uniquely determines n and the m_ρ.

Proof To prove existence, let $y_1, \ldots, y_N$ be any generators of G, and let G^* be the free abelian group on the N generators $Y_1, \ldots, Y_N$. There is a natural epimorphism $G^* \to G$ obtained by mapping each Y_ν into the corresponding y_ν; let H^* be its kernel. Since G^* is torsion-free, we can apply Lemma A2 to G^* and H^*, obtaining a base $X_1, \ldots, X_N$ for G^* and integers $M_1, \ldots, M_R$; let x_ν be the image of X_ν in G. The x_ν generate G, and $b_1 x_1 + \cdots + b_N x_N = 0$ if and only if $M_\rho | b_\rho$ for each $\rho \leqslant r$ and $b_\rho = 0$ for each $\rho > r$. Thus we have achieved all our claims except the statement that $M_1 > 1$. But if some $M_\rho = 1$ the corresponding $x_\rho = 0$, so that x_ρ can be deleted from the base for G which we have just constructed.

To prove uniqueness, suppose that we have a second such representation, given by dashed letters. We first prove that $n = n'$. For if not, let $n > n'$ and let p be a prime dividing m_1. Using the undashed representation we have an obvious epimorphism from G to the n-dimensional vector space over $\mathbf{F}_p$; hence this space must be generated by the images of the x_ν'. This is absurd because the set which they generate contains at most $p^{n'} < p^n$ elements.

Now for any $m > 0$ consider the group mG consisting of all mx with x in G. This has the representation in the theorem if we replace the x_ν by $m x_\nu$ and the m_ν by $m_\nu / (m_\nu, m)$, where the $m x_\nu$ for which $m_\nu | m$ must be deleted. Hence m_ρ is invariantly defined by the property that m_ρ is the least m for which a canonical representation of mG uses at most $n - \rho$ generators. □

Corollary *Any subgroup of a finitely generated abelian group G is finitely generated.*

Proof As in the proof of the theorem, construct a finitely generated torsion-free group G^* and an epimorphism $G^* \to G$. If H is a subgroup of G, let H^* be the inverse image of H. Then H^* is finitely generated, by Lemma A2, and the images of the generators of H^* generate H. □

Now let V be $\mathbf{Q}^n$ or $\mathbf{R}^n$ with the standard topology, and write $k = \mathbf{Q}$ or $\mathbf{R}$ respectively. A **lattice** Λ in V is a $\mathbf{Z}$-module which satisfies any two of the following three conditions.

(i) Λ spans V as a k-vector space.

(ii) Λ is discrete in the topological space V.

(iii) Λ is a free $\mathbf{Z}$-module on n generators.

Lemma A3 *Any two of these three properties imply the third.*

Proof Suppose first that any n elements of Λ are linearly dependent over k. If S is a maximal set of linearly independent elements of Λ containing $r < n$ elements, then the k-vector space generated by S contains Λ; but since it has dimension r it is not the whole of V, so (i) does not hold.

We now use the pigeonhole principle. Let $x_1, \ldots, x_n$ be elements of Λ linearly independent over k, so that the x_ν span V; and let Λ_0 be the $\mathbf{Z}$-module spanned by the x_ν. Suppose that Λ is discrete in V; thus there is an integer $M > 0$ such that the only element $\sum \lambda_\nu x_\nu$ of V with each $|\lambda_\nu| < M^{-1}$ which lies in Λ is the origin. The M^n boxes of the form $m_\nu/M \leqslant \lambda_\nu < (m_\nu + 1)/M$, where the m_ν are integers with $0 \leqslant m_\nu < M$, cover the cube $\mathcal{C}$ defined by $0 \leqslant \lambda_\nu < 1$ for each ν. Let $y_1, \ldots, y_N$ be representatives of distinct cosets of Λ/Λ_0; by translation we can assume that they all lie in the cube $\mathcal{C}$. Hence $N \leqslant M^n$, for otherwise y_r, y_s would lie in the same box and $y_r - y_s$ would be an element of Λ lying in the forbidden neighbourhood of the origin. Hence Λ is finitely generated (by the x_ν and y_i), and Λ is a free $\mathbf{Z}$-module on n generators by the last sentence of Lemma A2 with $G = \Lambda$ and $H = \Lambda_0$. We have therefore shown that (i) and (ii) imply (iii).

If (i) and (iii) hold, let $\{x_1, \ldots, x_n\}$ be a base for Λ; then $x_1, \ldots, x_n$ span V and any point of V can be written as $\sum \lambda_\nu x_\nu$ with λ_ν in k. The neighbourhood of the origin given by $|\lambda_\nu| < 1$ for each ν contains no point of Λ other than the origin, so Λ is discrete in V.

Finally, if (ii) and (iii) hold but (i) does not, let W be the k-vector space spanned by Λ. Then (i) and (ii) hold with W in place of V; so (iii) holds with $\dim W$ in place of n, which is a contradiction. $\qquad\square$

A1.2 Norms and Traces

Let K, k be fields with $K \supset k$ and $[K : k] = n$ finite, and let $x_1, \dots, x_n$ be a base for K as a k-vector space. For any y in K the endomorphism of the k-vector space K given by $x \mapsto yx$ has a characteristic polynomial $F(Y)$. Explicitly, there exist a_{ij} in k such that

$$yx_i = \sum a_{ij}x_j \qquad (A1)$$

and the **characteristic polynomial** of y is $F(Y) = \det(YI - A)$ where A is the matrix of the a_{ij}. A change of base takes A into $T^{-1}AT$ and hence does not affect F; and $F(y) = 0$ follows by regarding (A1) as equations for the x_i. We are particularly interested in the second and last coefficients of $\det(YI + A)$, which define the **trace** and **norm** of y:

$$\mathrm{Tr}_{K/k}(y) = \mathrm{trace}(A), \quad \mathrm{norm}_{K/k}(y) = \det(A).$$

If A_1, A_2 correspond to y_1, y_2 respectively then $A_1 \pm A_2$ corresponds to $y_1 \pm y_2$ and A_1A_2 to y_1y_2; so we obtain

$$\mathrm{Tr}(y_1 \pm y_2) = \mathrm{Tr}\,y_1 \pm \mathrm{Tr}\,y_2, \quad \mathrm{norm}(y_1y_2) = (\mathrm{norm}\,y_1)(\mathrm{norm}\,y_2). \qquad (A2)$$

Lemma A4 *Let $K \supset k$ with $[K : k]$ finite, let y be an element of K and let F, f be the characteristic polynomials of y for K/k and for $k(y)/k$. Then $F = f^m$ where $m = [K : k(y)]$.*

Proof Let $x_1, \dots, x_m$ be a base for K over $k(y)$. If $[k(y) : k] = r$ then $1, y, \dots, y^{r-1}$ are a base for $k(y)$ over k, and therefore the $x_\mu y^\rho$ form a base for K over k. But if A, B are the matrices whose characteristic polynomials are F, f respectively, then A consists of m copies of B down the main diagonal, with zeros everywhere else. $\qquad\square$

Corollary *Let $y_1, \dots, y_n$ in the algebraic closure $\bar{k}$ of k be the distinct conjugates of y over k. Then $[K : k] = ns$ for some integer s, and*

$$\mathrm{norm}_{K/k}y = (y_1 \cdots y_n)^s, \quad \mathrm{Tr}_{K/k}y = s(y_1 + \cdots + y_n). \qquad (A3)$$

If $\sigma_1, \ldots, \sigma_N$ are the distinct embeddings of K into $\bar{k}$ which fix k element-wise, then $[K : k] = NS$ for some integer S, and

$$\text{norm}_{K/k}y = \prod(\sigma_\nu y)^S, \quad \text{Tr}_{K/k}y = S\sum(\sigma_\nu y).$$

If $k(y)$ is separable over k then $s = [K : k(y)]$; if K is separable over k then $S = 1$.

Proof The equations (A3) follow from the fact that the roots of f are the y_ν, possibly repeated, and they have equal multiplicity; if $k(y)$ is separable over k then this multiplicity is 1 and $n = [k(y) : k]$. For each y_ν there is a homomorphism $k(y) \to \bar{k}$ given by $y \mapsto y_\nu$, and each of these can be extended to K in the same number of ways. $\qquad\square$

We could have defined norm and Trace by the formulae in this corollary — and this is what we had to do when we defined the norm of an ideal in §4. The crucial formulae (A2) follow at once and we do not need to mention the characteristic polynomial. But we would instead have to prove that the results lie in k — though this is trivial in the separable case, which is all that concerns us in this book.

Theorem A2 *Let $K \supset L \supset k$ be a tower of extensions, with $[K : k]$ finite, and let y be an element of K. Then*

$$\text{norm}_{L/k}(\text{norm}_{K/L}y) = \text{norm}_{K/k}y, \quad \text{Tr}_{L/k}(\text{Tr}_{K/L}y) = \text{Tr}_{K/k}y.$$

Proof We use the notation of the last corollary. In the first equation, each side is the product of the y_ν each taken the same number of times; and this number is determined by the fact that each side is the product of $[K : k]$ conjugates of y. For the second equation we need only replace 'product' by 'sum'. $\qquad\square$

A1.3 Haar measure

A **topological group** is a group which, as a set, is equipped with a topology such that the group operations (multiplication and inversion) are continuous. There is one major theorem about topological groups; and every serious mathematician needs to know the statement of it though not the proof.

Theorem A3 *Let G be a locally compact topological group. There exists on G a measure μ, unique up to multiplication by a constant, such that* *every compact subset of G is measurable and*

$$\mu(g_0 S) = \mu(S)$$

for any g_0 in G and any measurable subset S of G. There is a corresponding integral $\int_G f(g)d\mu$ with the property

$$\int_G f(g_0 g)d\mu = \int_{g_0 G} f(g)d\mu.$$

These are called the **left Haar measure** and **left Haar integral** respectively. There is a similar theorem for the right Haar measure and right Haar integral. Left and right Haar measure are frequently the same (for example if G is either commutative or compact); but this is not always so.

The proof of this theorem is difficult; but one really uses it only as motivation. In the cases which occur in practice one usually needs to know the Haar measure explicitly and it is usually obvious what it is; but one hardly ever needs the uniqueness assertion.

It is sometimes obvious how one should normalise the Haar measure; for example, if G is compact or has a particularly important compact subgroup G_0 then it is natural to choose μ so that $\mu(G_0) = 1$. Again, the natural normalisation of μ on $\mathbf{R}$ is the one which induces $\mu(\mathbf{R}/\mathbf{Z}) = 1$; but there is no obviously best normalisation of the Haar measure on $\mathbf{C}$.

Now let K be a locally compact topological field. Associated with K there are two topological groups — K with the addition law and K^* with the multiplication law — and the corresponding Haar measures μ^+ and $\mu^\times$ are different. These are respectively the **additive** and the **multiplicative** Haar measures on K. Let α be any non-zero element of K. The map

$$S \mapsto \mu^+(\alpha S)$$

is an additive Haar measure on K, so it has the form $S \mapsto c_\alpha \mu^+(S)$ for some c_α by the uniqueness of Haar measure. One frequently writes $c_\alpha = \|\alpha\|$, thus defining a function from K^* to the positive reals; note that $\|\alpha\|$ does not depend on the choice of μ^+. (By convention we write $\|0\| = 0$.) Clearly

$$\|\alpha_1 \alpha_2\| = \|\alpha_1\| \cdot \|\alpha_2\|;$$

and though $\|.\|$ need not induce a metric on K, under modest extra conditions it induces a topology on K which is just the one we began with. Moreover

$$\int f(x)d\mu^\times = \int f(x)\|x\|^{-1}d\mu^+ \qquad \text{(A4)}$$

if we adjust the constants in $\mu^\times, \mu^+$ suitably.

The simplest cases of Haar measure are given by (i) to (iii) below; (iv) and (v) are crucial for applications to algebraic number theory.

(i) If G is a finite group, or more generally if G has the discrete topology, $\mu(\mathcal{S})$ is just the number of elements in $\mathcal{S}$. If K is a finite field then $\|\alpha\| = 1$ for $\alpha \neq 0$.

(ii) $\mathbf{R}$ is a field and the Haar measure on $\mathbf{R}$ is the standard one. Thus $\|x\| = |x|$ and the Haar measure on $\mathbf{R}^*$ which is given by (A4) is $\mathcal{S} \mapsto \int_{\mathcal{S}} |x|^{-1} dx$.

(iii) $\mathbf{C}$ is also a field. One possible Haar measure on $\mathbf{C}$ is given by area in the complex plane, but it is customary to double this, giving the measure associated with $dz \wedge d\bar{z}$. In either case $\|z\| = |z|^2$; this does not give a metric, but $\|z\|^{1/2}$ does. Again the Haar measure on $\mathbf{C}^*$ is given by (A4).

(iv) Let k be an algebraic number field and $\mathfrak{p}$ a prime ideal of k; then $k_\mathfrak{p}$ is a locally compact field. We have $\|\alpha\|_\mathfrak{p} = (\text{Norm}\,\mathfrak{p})^{-m}$ if α is in k^* and $\mathfrak{p}^m\|\alpha$. Every residue class $\bmod\,\mathfrak{p}^m$ for fixed $m \geqslant 0$ has measure $\mu^+(\mathfrak{o}_\mathfrak{p})(\text{Norm}\,\mathfrak{p})^{-m}$. It might appear natural to normalise the additive Haar measure so that $\mu^+(\mathfrak{o}_\mathfrak{p}) = 1$, and this is what some writers do; but following Tate we shall normalize it so that $\mu^+(\mathfrak{o}_\mathfrak{p}) = (\text{Norm}\,\mathfrak{d}_\mathfrak{p})^{-1/2}$ where $\mathfrak{d}_\mathfrak{p}$ is the local different introduced in §8. The benefits of this will become clear in §A2.2 and in the Corollary to Lemma 26. The measure on V_k will be the product of the measures on the k_v.

(v) There are also several plausible ways to normalize the measure on $k_\mathfrak{p}^*$. We choose to require $\mu^\times(\mathfrak{o}_\mathfrak{p}^*) = 1$, and to take the measure on J_k to be the product of the measures on the k_v^*. However, some writers prefer to take $\mu^\times(\mathfrak{o}_\mathfrak{p}^*) = (\text{Norm}\,\mathfrak{p} - 1)/(\text{Norm}\,\mathfrak{p})$; if so, the set consisting of the units congruent to $\beta \bmod \mathfrak{p}^m$ for any fixed unit β and any $m > 0$ has measure $(\text{Norm}\,\mathfrak{p})^{-m}$. For those who also take $\mu^+(\mathfrak{o}_\mathfrak{p}) = 1$, the relation (A4) is preserved.

A2 Additional topics

This section falls into two parts. The first two subsections outline the theory of Fourier transforms on locally compact abelian groups, which is an essential foundation for §15. The third one is not used anywhere in this book, except for a passing reference at the end of §18; but I hope that the reader who pursues the advanced theory will find it useful.

A2.1 Characters and duality

Denote by $\mathbf{T}$ the **circle group** $|z| = 1$ with the usual topology, and let G be any locally compact abelian group; then a **character** of G is defined to be a continuous homomorphism $c : G \to \mathbf{T}$. The following trivial remark will be used repeatedly.

Lemma A5 *Let* χ *be a character on a compact group* G; *then*

$$\int_G \chi(x)\,d\mu_x = \left\{ \begin{array}{ll} \mu(G) & \text{if } \chi \text{ is trivial on } G, \\ 0 & \text{otherwise.} \end{array} \right.$$

Proof We need only prove the second statement. Choose x_0 in G with $\chi(x_0) \neq 1$ and write xx_0 for x in the integral; then

$$\int_G \chi(x)\,d\mu = \int_G \chi(xx_0)\,d\mu = \chi(x_0) \int_G \chi(x)\,d\mu,$$

from which the statement follows. □

The characters of G form a multiplicative group, normally denoted by $\widehat{G}$; to write them additively, we use the identification $\mathbf{T} \approx \mathbf{R}/\mathbf{Z}$. We can topologize $\widehat{G}$ as follows. Fix a character c_0 of G; then a base for the open neighbourhoods of c_0 in $\widehat{G}$ is given by the set of characters c on G such that

$$|c(g) - c_0(g)| < \epsilon \text{ for } g \text{ in } \mathcal{S}$$

where $\epsilon > 0$ and $\mathcal{S}$ is any compact subset of G. It is easy to check that with this topology $\widehat{G}$ becomes a locally compact topological group and each element of G induces a character on $\widehat{G}$. But much more is true:

Theorem A4 (Pontryagin duality) *We can identify the character group of* $\widehat{G}$ *with* G. *If* G *is compact then* $\widehat{G}$ *is discrete; if* G *is discrete then* $\widehat{G}$ *is compact. If* H *is a closed subgroup of* G *and* $H^\sharp$ *consists of those elements of* $\widehat{G}$ *which are trivial on* H, *then* $H^\sharp$ *is closed in* $\widehat{G}$ *and there are canonical isomorphisms* $\widehat{H} \approx \widehat{G}/H^\sharp$ *and* $\widehat{G/H} \approx H^\sharp$.

Corollary *Any character on a closed subgroup of* G *can be extended (non-uniquely) to the whole of* G.

Note that if the G_i are subgroups of G such that any neighbourhood of the identity in G contains some G_i then for any character c on G we can find some G_i on which c is trivial. For let $\mathcal{N}$ be the neighbourhood of 1 in $\mathbf{T}$ given by $\Re z > \frac{1}{2}$; then $c^{-1}\mathcal{N}$ is a neighbourhood of the identity in G and therefore contains some G_i. But $c(G_i)$ is a subgroup of $\mathbf{T}$ contained in $\mathcal{N}$, and must therefore be trivial because $\mathcal{N}$ contains no non-trivial subgroups. A set of such G_i is called a set of **small subgroups**.

We now find the character groups of some of the topological groups we shall encounter. All these calculations can be done in an elementary fashion, so that logically we shall not need Theorem A4.

Consider first $\mathbf{Z}$: now we can choose $c(1)$ arbitrarily and define c by $c(n) = c(1)^n$, and any such c is continuous. Hence $\widehat{\mathbf{Z}} = \mathbf{T}$ and by duality $\widehat{\mathbf{T}} = \mathbf{Z}$; it is easy to check directly that any character on $\mathbf{T}$ has the form $z \mapsto z^n$ for some n in $\mathbf{Z}$ by considering its kernel, which must be either the whole of $\mathbf{T}$ or a finite subgroup. Since $\mathfrak{o}_k \approx \mathbf{Z}^n$, its dual is $\widehat{\mathfrak{o}_k} \approx \mathbf{T}^n$.

Let k be an algebraic number field. In view of Lemma 26 we expect to give $\mathbf{Q}$ or k the discrete topology; and in fact there are no other sensible topologies which make these groups locally compact. We are not interested in all their characters, but only in those that extend to some completion; and these are best obtained by finding directly the characters on the possible completions. Let k_v be any completion of k and let $\mathbf{Q}_w$ be the closure of $\mathbf{Q}$ within k_v, so that w is the valuation on $\mathbf{Q}$ which is the restriction of v. $\mathbf{Q}_w$ must be either $\mathbf{R}$ or $\mathbf{Q}_p$. In either case there are natural maps $\iota_w : \mathbf{Q}_w \to \mathbf{R}/\mathbf{Z}$, being given for $\mathbf{Q}_w = \mathbf{Q}_p$ by $\mathbf{Q}_p \to \mathbf{Q}_p/\mathbf{Z}_p \to \mathbf{R}/\mathbf{Z}$ because the elements of $\mathbf{Q}_p/\mathbf{Z}_p$ can be identified with those elements of $\mathbf{Q}/\mathbf{Z}$ whose denominators are powers of p. Using ι_w we can define the function $x \mapsto \exp(2\pi i x)$ for any x in $\mathbf{Q}_w$. In the following lemma the signs have been chosen so as to simplify the corresponding statement in the adèlic case.

Lemma A6 *We can identify $\widehat{k_v}$ with k_v, the character c_α corresponding to α in k_v being given by*

$$c_\alpha(\beta) = \exp(\pm 2\pi i \operatorname{Tr}_{k_v/\mathbf{Q}_w}(\alpha\beta)),$$

where the sign is chosen to be minus for Archimedean and plus for non-Archimedean places.

Proof Most of this assertion is easy. With the definition above, c_α is a character; for it is a homomorphism, and it is continuous because Trace is so. The c_α form a group algebraically isomorphic to k_v under $\alpha \mapsto c_\alpha$, for c_α trivial implies that $\operatorname{Tr}(\alpha\beta)$ is in $\mathbf{Z}$ for all β, which is only possible if $\alpha = 0$. Moreover the group of c_α is homeomorphic to k_v under this correspondence. For let $\mathcal{S}_M$ be the compact set $\|\beta\|_v \leqslant M$ where M is fixed; then $\|\alpha\|_v$ small implies $\operatorname{Tr}(\alpha\beta)$ small uniformly in β and hence c_α

is close to the identity character. Thus $\alpha \mapsto c_\alpha$ is continuous. Conversely, suppose that c_α is close to the identity character. If w is Archimedean this means that $|\operatorname{Tr}(\alpha\beta) - n_\beta| < \epsilon$ for each β in S_M and some n_β in $\mathbf{Z}$, where we require $\epsilon \leqslant \frac{1}{4}$. If n_β is ever non-zero then by replacing β by βx where $|x| < 1$ and $x n_\beta$ is not close to an integer, we obtain a contradiction. So $|\operatorname{Tr}(\alpha\beta)| < \epsilon$ for all β in S_M, and this implies that α is small with ϵ. If instead w corresponds to p then $|\operatorname{Tr}(\alpha\beta) - n_\beta| < \epsilon$ where this time n_β is in $\mathbf{Z}_p$ and we shall require $\epsilon < \frac{1}{4p}$; if $\operatorname{Tr}(\alpha\beta)$ is ever not in $\mathbf{Z}_p$ then by replacing β by $p^m \beta$ for suitably chosen m we obtain a contradiction. Now take M large; if c_α is close enough to the identity character, $\alpha\beta$ must be in $\mathfrak{d}^{-1}$ for all β with $\|\beta\|_v \leqslant M$, where $\mathfrak{d}$ is the different, and this implies that $\|\alpha\|_v$ is small.

It remains to show that we have found all the characters of k_v. But the characters which we have found certainly form a closed subgroup $H^\sharp$ in the notation of Theorem A4; and if this is not the whole of k_v then there is a non-trivial subgroup $H \subset k_v$ on which $H^\sharp$ is trivial. Choosing $\beta_0 \neq 0$ in H we obtain a contradiction. $\qquad\square$

Consider next $\widehat{V_k}$. Let S be any finite set of places, including all the Archimedean places, and let G_S be the subgroup of V_k consisting of those adèles $\prod \alpha_v$ for which $\alpha_v = 0$ for v in S and α_v is in $\mathfrak{o}_v$ for all v. The G_S are a set of small subgroups in the sense of the remarks below the Corollary to Theorem A4, so any character c on V_k is trivial on some G_S, say G_{S_0}. Let c_v denote the character induced on k_v by the natural embedding $k_v \to V_k$; thus c_v is trivial on $\mathfrak{o}_v$ for almost all v. Since any β in V_k is the sum of an element in G_{S_0} and a finite set of elements of various k_v,

$$c(\beta) = \prod c_v(\beta_v),$$

where almost all the factors on the right are 1. Conversely, if we have any set of characters c_v on the k_v such that c_v is trivial on $\mathfrak{o}_v$ for almost all v, this formula defines a character on V_k. This determines the algebraic structure of $\widehat{V_k}$. As for its topology, any compact set in V_k is contained in $\prod S_v$ where each S_v is compact and almost every S_v is $\mathfrak{o}_v$; hence a base for the open neighbourhoods of the trivial character is given by the sets $\prod C_v$, where each C_v is an open neighbourhood of the trivial character in k_v and almost every C_v consists of all characters which are trivial on $\mathfrak{o}_v$. If v corresponds to a finite unramified prime $\mathfrak{p}$, such a character corresponds to an element of $\mathfrak{o}_v$ under the identification in Lemma A6. We have therefore proved

Lemma A7 *There is a natural identification $\widehat{V_k} \approx V_k$.*

Explicitly, the character c_α corresponding to $\alpha = \prod \alpha_v$ in V_k is given by

$$c_\alpha \left(\prod \beta_v \right) = \exp \left(2\pi i \sum (\pm \iota_w \mathrm{Tr}_{k_v/\mathbf{Q}_w}(\alpha_v \beta_v)) \right) \tag{A5}$$

where as before w is the valuation of $\mathbf{Q}$ which is the restriction of v, and ι_w maps $\mathbf{Q}_w$ to $\mathbf{R}/\mathbf{Z}$. As in Lemma A6 the sign is plus for non-Archimedean and minus for Archimedean places. One useful consequence is that if α, β are both in V_k, then

$$c_\alpha(\beta) = c_1(\alpha\beta). \tag{A6}$$

Corollary *The character c_α is trivial on k if and only if α is in k.*

Proof Suppose first that α, β are both in k. By (39),

$$\iota_\infty \sum \mathrm{Tr}_{k_v/\mathbf{Q}_\infty}(\alpha\beta) = \sum_p \iota_p \sum_{\mathfrak{p}|p} \mathrm{Tr}_{k_\mathfrak{p}/\mathbf{Q}_p}(\alpha\beta)$$

where the sum on the left is taken over all Archimedean places; for each side is equal to the image of $\mathrm{Tr}_{k/\mathbf{Q}}(\alpha\beta)$ in $\mathbf{Q}/\mathbf{Z}$. Thus the sum on the right of (A5) is 0. (It was for this reason that we introduced the unnatural looking sign convention in Lemma A6.) Now let G be the additive group of all adèles $\alpha = \prod \alpha_v$ such that c_α is trivial on k. We have shown that $G \supset k$, and G is a k-vector space; thus G/k is a k-vector space which is a subspace of V_k/k. But the latter is compact, by Lemma 26, so G/k is trivial. $\qquad\square$

We shall not need to know $\widehat{J_k}$ in what follows; and this is just as well because in general no satisfactory description of it is known. But we shall need some information about its **quasi-characters** — that is to say, its continuous homomorphisms into $\mathbf{C}^*$. In particular if χ is a quasi-character so is $\alpha \mapsto \|\alpha\|^s \chi(\alpha)$, where the function $\|.\|$ is that defined after the proof of Lemma 27.

A2.2 Fourier transforms

Let $f(g)$ be a complex-valued function on a locally compact additive group G, and let μ be a Haar measure on G. Suppose that f is continuous and integrable on G. The **Fourier transform** of f is the function on $\widehat{G}$ given by

$$\hat{f}(c) = \int_G f(g)\overline{c(g)}d\mu.$$

The complex conjugacy sign appears here for historical reasons; however, complex conjugacy is bound to appear somewhere in the theory. The importance of this definition comes from the duality relation:

Theorem A5 *There is a Haar measure* $\hat{\mu}$ *on* $\widehat{G}$ *such that*

$$f(g) = \hat{\hat{f}}(-g) = \int_{\widehat{G}} \hat{f}(c)c(g)d\hat{\mu}$$

provided that the integral exists.

Since $\widehat{G}$ only determines the Haar measure on it up to multiplication, to find the correct $\hat{\mu}$ for any particular G we need to verify the result for one particular function f. If we can identify G with $\widehat{G}$ then we can normalize μ; for if we multiply μ by a constant we have to divide $\hat{\mu}$ by the same constant, and there is a unique positive multiplier for which $\hat{\mu}$ is equal to μ. We shall show below that if G is $\mathbf{R}, \mathbf{C}$ or $k_{\mathbf{p}}$ then the measure μ which was chosen in §A1.3 is the one which achieves this. It will then follow that the same statement holds for V_k.

The modern theory of Fourier transforms recognizably includes the classical analytic theory — though the classical results were proved under weaker conditions on f. There are three important special cases in classical pure mathematics: Fourier series, Fourier integrals, and Mellin transforms. The first two are straightforward. Since $\widehat{\mathbf{Z}} = \mathbf{T} \approx \mathbf{R}/\mathbf{Z}$, Theorem A5 gives

$$f(x) = \sum_{-\infty}^{\infty} c_n e^{2\pi i n x} \quad \Longleftrightarrow \quad c_n = \int_0^1 f(x)e^{-2\pi i n x}dx.$$

Next consider $\widehat{\mathbf{R}} \approx \mathbf{R}$; then

$$\hat{f}(y) = \int_{-\infty}^{\infty} f(x)e^{2\pi i x y}dx \quad \Longleftrightarrow \quad f(x) = \int_{-\infty}^{\infty} \hat{f}(y)e^{-2\pi i x y}dy. \quad \text{(A7)}$$

Thus in this case $\mu = \hat{\mu}$. The easiest way to check that the multiplicative constant is correct is to note that $f(x) = e^{-\pi x^2}$ implies $\hat{f}(y) = e^{-\pi y^2}$.

Classically, the Fourier duality formula for $\mathbf{C}$ was not important enough to warrant a name. It asserts that if

$$\hat{f}(w) = \iint f(z)\exp(2\pi i\Re(zw))\,dz \wedge d\bar{z}$$

then (at least up to a constant factor)

$$f(z) = \iint \hat{f}(w)\exp(-2\pi i\Re(zw))\,dw \wedge d\bar{w}.$$

To check that the factor here is correct, take $f(z) = g_1(x\sqrt{2})g_2(y\sqrt{2})$ where $z = x + iy$; if $w = u + iv$ then $\hat{f}(w) = \hat{g}_1(u\sqrt{2})\hat{g}_2(-v\sqrt{2})$ where $\hat{g}_1, \hat{g}_2$ are given by (A7).

To check that $\hat{\mu} = \mu$ for $k_{\mathfrak{p}}$ take

$$f(x) = \begin{cases} 1 \text{ for } x \text{ in } \mathfrak{o}_{\mathfrak{p}}, \\ 0 \text{ otherwise}; \end{cases}$$

then by Lemma A5

$$\hat{f}(y) = \int_{\mathfrak{o}_{\mathfrak{p}}} \exp(-2\pi i \operatorname{Tr}_{k_{\mathfrak{p}}/\mathbf{Q}_p}(xy)) \, d\mu_x^+$$
$$= \begin{cases} (\operatorname{Norm} \mathfrak{d}_{\mathfrak{p}})^{-1/2} & \text{for } y \text{ in } \mathfrak{d}_{\mathfrak{p}}^{-1}, \\ 0 & \text{otherwise}, \end{cases}$$

because the integrand in the middle expression is a character on $\mathfrak{o}_{\mathfrak{p}}$. Similarly we obtain

$$\int_{k_{\mathfrak{p}}} \hat{f}(y) \exp(2\pi i \operatorname{Tr}_{k_{\mathfrak{p}}/\mathbf{Q}_p}(xy)) \, d\mu_y^+ = f(x)$$

since $\mu^+(\mathfrak{d}_{\mathfrak{p}}^{-1}) = (\operatorname{Norm} \mathfrak{d}_{\mathfrak{p}})\mu^+(\mathfrak{o}_{\mathfrak{p}})$. Thus $\hat{\mu} = \mu$.

The Mellin transform formula is derived from (A7) exactly as in the classical theory. For let $g(z) = \sum_1^\infty a_n \exp(2\pi i n z)$ where the a_n are complex numbers such that $a_n = O(n^{\sigma-1-\epsilon})$ for some real σ and some $\epsilon > 0$; this series is absolutely convergent in the upper half-plane $\Re z > 0$, and $g(z)$ is periodic with period 1. If $L(s) = \sum a_n n^{-s}$ where $s = \sigma + i\tau$ then

$$\Gamma(s)(2\pi)^{-s}L(s) = \sum a_n \int_0^\infty e^{-2\pi n y} y^{s-1} dy = \int_0^\infty g(iy) y^{s-1} dy.$$

But this simply says that the left hand side, considered as a function of τ, is the Fourier integral transform of $2\pi e^{2\pi x \sigma} g(ie^{2\pi x})$. After some tidying up, the dual formula becomes

$$g(iy) = \frac{1}{2\pi i} \int \Gamma(s)(2\pi y)^{-s} L(s) ds$$

where the integral is taken along the vertical line $\Re s = \sigma$.

Theorem A6 (Poisson Summation Formula) *Let H be a discrete subgroup of G such that G/H is compact, and let $H^\sharp$ consist of the elements of $\widehat{G}$ which are trivial on H, so that $H^\sharp = \widehat{G/H}$. Then*

$$\mu(G/H) \sum_H f(h) = \sum_{H^\sharp} \hat{f}(h^\sharp) \tag{A8}$$

provided that f *is integrable on* G, $\sum_H f(g+h)$ *is absolutely convergent uniformly in* g *and* $\sum \hat{f}(h^\sharp)$ *is absolutely convergent.*

Proof $H^\sharp$ is discrete and $\widehat{G}/H^\sharp$ is compact, by Theorem A4. Define the function $\phi(x)$ on G/H by $\phi(x) = \sum_H f(x+h)$; then

$$\int_{G/H} \phi(x)d\mu = \int_G f(g)d\mu, \quad \hat{\phi}(h^\sharp) = \int_{G/H} \phi(x)\overline{h^\sharp(x)}d\mu. \tag{A9}$$

Now Theorem A5 gives

$$\phi(x)\mu(G/H) = \sum_{H^\sharp} h^\sharp(x)\hat{\phi}(h^\sharp)$$

up to a constant factor. To see that the constant is correct, set $\phi(x) = 1$; then $\hat{\phi}(1) = \mu(G/H)$ and $\hat{\phi}(h^\sharp) = 0$ otherwise, the latter result coming from writing xx_0 for x in the second equation (A9) where $h^\sharp(x_0) \neq 1$. Also

$$\hat{\phi}(h^\sharp) = \int_{G/H} \phi(x)\overline{h^\sharp(x)}d\mu = \int_G f(g)\overline{h^\sharp(g)}d\mu = \hat{f}(h^\sharp),$$

the change in the order of summation and integration being justified by the hypotheses on f. Hence

$$\phi(x)\mu(G/H) = \sum h^\sharp(x)\hat{f}(h^\sharp),$$

and writing $x = 0$, $\phi(0) = \sum f(h)$ gives the theorem. $\qquad\square$

If we multiply μ by a constant, we multiply both sides of (A8) by the same constant; so the theorem is true for any choice of μ. The special case which will be needed in §15 is as follows.

Corollary *Let* $f(\xi)$ *be a function on* V_k *which is integrable and for which* $\sum_{x \text{ in } k} f(\alpha(x+\xi))$ *is absolutely convergent for all adèles* ξ *and idèles* α, *uniformly in* ξ *for each* α. *Suppose also that* $\sum_{x \text{ in } k} \hat{f}(\alpha x)$ *converges for each* α, *where we have identified* $\widehat{V_k}$ *with* V_k. *Then*

$$\mu(V_k/k)\|\alpha\| \sum_{x \text{ in } k} f(\alpha x) = \sum_{x \text{ in } k} \hat{f}(\alpha^{-1}x).$$

Proof Write $g(\xi) = f(\alpha\xi)$; then

$$\hat{g}(\eta) = \int_V g(\xi)\overline{c_\eta(\xi)}d\mu = \int_V f(\alpha\xi)\overline{c_1(\xi\eta)}d\mu$$

$$= \|\alpha\|^{-1}\int_V f(\xi)\overline{c_1(\alpha^{-1}\xi\eta)}d\mu = \|\alpha\|^{-1}\hat{f}(\alpha^{-1}\eta),$$

where to go from the first line to the second we have written $\alpha^{-1}\xi$ for ξ. Now apply the theorem to $g(\xi)$ with $G = V_k$ and $H = k$, and use the fact that $H^\sharp = k$ by the Corollary to Lemma A7.   □

If we identify $\widehat{V_k}$ with V_k, the symmetry property of the Fourier transform becomes

$$\hat{f}(\eta) = \int_V f(\xi)\overline{c_1(\xi\eta)}d\mu \quad \Longleftrightarrow \quad f(\xi) = \int_V \hat{f}(\eta)c_1(\xi\eta)d\mu \qquad (A10)$$

where μ is normalized by the condition $\mu(V_k/k) = 1$. For we know from Theorem A5 and (A6) that $f(\xi) = A\int_V \hat{f}(\eta)c_1(\xi\eta)d\mu$ for some constant A depending on μ. Applying Theorem A6 to both f and $\hat{f}$ and remembering that $\overline{c_1(\xi\eta)} = c_1(-\xi\eta)$ we obtain $A(\mu(V_k/k))^2 = 1$; so $A = 1$ is equivalent to $\mu(V_k/k) = 1$.

A2.3 Galois theory for infinite extensions

Let K be a separable normal extension of a field k, with Galois group G. If $[K : k]$ is finite, the fundamental theorem of Galois theory states that there is a one-one correspondence between fields L with $K \supset L \supset k$ and groups $H \subset G$, given by the relations

L consists of those elements of K which are fixed under H,

H consists of those elements of G which fix L elementwise.

The usual proofs of this depend on the finiteness of $[K : k]$, because they use counting arguments; so they collapse when $[K : k]$ is infinite. It is natural to ask how much the conclusion needs to be changed in that case.

Lemma A8 *Let K, L be fields such that $K \supset L \supset k$ and K is Galois over k. If H is the Galois group of K over L, then the fixed field of H is L.*

Proof This follows directly from the finite case. For let ξ be any element of K not in L and let $\mathfrak{K}$ with $K \supset \mathfrak{K} \supset L$ be the splitting field of the minimal polynomial of ξ over L. There is an L-automorphism of $\mathfrak{K}$ which does not leave ξ fixed, and this can be extended to an L-automorphism σ of K with the help of Zorn's Lemma. Since σ is an element of H, the fixed field of H cannot be larger than L.   □

There are in general too many subgroups H of G for them all to be Galois groups of K/L for some intermediate field L; so how do we identify those H which have this property? To answer this, we introduce a topology on G.

Let K_i run through those subfields of K which are normal over k and for which $[K_i : k]$ is finite, and let

$$G \supset G_i = \mathrm{Gal}(K/K_i), \quad S_i = G/G_i = \mathrm{Gal}(K_i/k).$$

If $K_i \supset K_j$ there is a natural homomorphism $\phi_{ij} : S_i \to S_j$. The S_i form an inverse system of groups with connecting homomorphisms ϕ_{ij}, whose inverse limit is G with the obvious maps $\phi_i : G \to S_i = G/G_i$. For let σ be any element of G. Each ξ in K is in some K_i; and the actions of σ and $\phi_i\sigma$ on ξ are the same. Hence σ is uniquely determined by a knowledge of all the $\phi_i\sigma$.

The S_i are finite groups, so we give each of them the discrete topology; and we give G the topology induced on it as an inverse limit — that is, the finest topology such that each ϕ_i is continuous. A base for the open subsets of G is given by the cosets of the G_i, and any such set is also closed because it is the complement of the union of the other cosets of the same G_i. The topology obtained in this way is Hausdorff, and with it G is compact. However, it is not known, even when $G = \mathrm{Gal}(\overline{\mathbf{Q}}/\mathbf{Q})$, whether every subgroup of finite index in G is open; this is one of a number of related unsolved questions, all of which appear to be very difficult.

Lemma A9 *Let H be a subgroup of G. Then H is the Galois group of K over some field L with $K \supset L \supset k$ if and only if H is topologically closed.*

Proof Suppose first that $H = \mathrm{Gal}(K/L)$. If σ is in the closure of H and ξ is any element of L, we have to prove that $\sigma\xi = \xi$. Let $K_0 \subset K$ be the splitting field for the minimal polynomial of ξ over k; then K_0 is Galois over k and $[K_0 : k]$ is finite. Let G_0, S_0 and ϕ_0 be as above. Now $\phi_0 H \subset S_0$ is closed because S_0 is discrete, and it leaves ξ fixed; hence σ belongs to the closed set $\phi_0^{-1}\phi_0 H$ and leaves ξ fixed.

Conversely, if H is closed let L be its fixed field. If σ is not in H, we have to show that there are elements of L which are not fixed by σ. There is a basic neighbourhood $\mathcal{N}$ of σ which does not meet H — that is, there is a K_1 finite and Galois over k with $\phi_1 : G \to S_1 = \mathrm{Gal}(K_1/k)$ such that $\phi_1 H$ does not contain $\phi_1\sigma$. Let $\mathfrak{K}_1 \subset K_1$ be the fixed field of $\phi_1 H \subset S_1$. Then $\mathfrak{K}_1 \subset L$ because $\mathfrak{K}_1$ is fixed elementwise by H; but $\mathfrak{K}_1$ is not fixed under σ. $\qquad\Box$

Putting the last two lemmas together, we obtain the Fundamental Theorem of Galois Theory for arbitrary extensions (which contains the corresponding one for finite extensions):

Theorem A7 *In the notation above, there is a one-one correspondence between fields L with $K \supset L \supset k$ and closed subgroups H of G.*

There is one case in which we know how to describe G explicitly: that is when k is a finite field $\mathbf{F}_q$ and $K = \bar{k}$ is its algebraic closure. Now there is just one extension of k of each finite degree n, which we call $K^{(n)}$. Since $S^{(n)} = \mathrm{Gal}(K^{(n)}/k)$ is cyclic of order n and is generated by the Frobenius endomorphism $\sigma : x \mapsto x^q$, we can identify $S^{(n)}$ with $\mathbf{Z}/(n)$ by identifying σ^ν with $\nu \bmod n$. Here $K^{(m)} \supset K^{(n)}$ if and only if $n|m$, and the map $\phi_{m,n}$ is then identified with the map $\mathbf{Z}/(m) \to \mathbf{Z}/(n)$ which sends $\nu \bmod m$ into $\nu \bmod n$.

If m, n are coprime, the degree of the compositum $K^{(m)} K^{(n)}$ must be a multiple of both m and n, whence $K^{(m)} K^{(n)} = K^{(mn)}$. The natural map $S^{(mn)} \to S^{(m)} \times S^{(n)}$ of Galois groups is one-one and can be identified with the natural map of residue groups; and it is an isomorphism. In forming G, the inverse limit of the $S^{(n)}$, we can therefore treat the different prime factors of the n separately; thus G is naturally isomorphic to the product of the inverse limits of the $S^{(p^\nu)} \approx \mathbf{Z}/(p^\nu)$. These inverse limits are just the $\mathbf{Z}_p$.

Exercises

Chapter 1

1.1 For the ring $R = \mathbf{C}[x, y]$, which has unique factorization, prove that

(i) apart from (0) and R itself, there are two kinds of prime ideals, one kind being maximal and the other not,

(ii) any non-zero ideal can be written in at most one way as a product of prime ideals,

(iii) there are ideals which cannot be written as a product of prime ideals,

(iv) division by ideals, even when possible, is not always unique.

[Think geometrically. The two kinds of prime ideals correspond to points and irreducible curves.]

1.2 Show that neither of $\mathbf{Z}[\sqrt{-6}]$ and $\mathbf{Z}[\sqrt{5}]$ is a unique factorization domain.

[Consider the factorization into irreducible elements of 6 in the first case and 4 in the second.]

1.3 Find a unit in $\mathbf{Q}(\sqrt[3]{6})$ and show that this field has class number $h = 1$.

1.4 Find a unit in $\mathbf{Q}(\sqrt[3]{22})$ and show that this field has class number $h = 3$. Deduce that $X^3 + 22Y^3 + 3Z^3 = 0$ has no rational solutions.

[If x, y, z is a solution with x, y coprime integers, show that $\mathfrak{p}_3$ is not principal, where $\mathfrak{p}_3 | 3$, but that $(x + y\sqrt[3]{22}) = \mathfrak{p}_3 \mathfrak{a}^3$. The equation is in fact soluble in every $\mathbf{Q}_v$, but to prove this requires ideas not in this book; see for example [Ca].]

1.5 Suppose that the extension $K/\mathbf{Q}$ is normal and has a Galois group which is simple but not cyclic. Show that there is no rational prime p such that (p) remains prime in K.

135

1.6 Show that the ring of all algebraic integers is not Noetherian, and deduce that $\overline{\mathbf{Q}}$ is a Noetherian ring not all of whose subrings are Noetherian.

[Fix a prime p and let M_n be the ideal generated by $p^{1/n}$; show that the M_n form a strictly increasing sequence.]

1.7 Suppose that k contains ζ, a primitive p-th root of unity where p is prime, and that K is Galois over k with $[K : k] = p$; and write $G = \mathrm{Gal}(K/k) \approx C_p$. Show that $K = k(\sqrt[p]{\alpha})$ for some α in k.

[Let σ be a generator of G. Take $\alpha = \sum_0^{p-1} \zeta^\nu \cdot \sigma^\nu \beta$ for β in K; and show that one can choose β so that $\alpha \neq 0$.]

1.8 Let K, k be algebraic number fields with K totally complex, k totally real and $[K : k] = 2$. If μ_K denotes the group of roots of unity in K, show that $[\mathfrak{O}_K^* : \mathfrak{o}_k^* \mu_K] = 1$ or 2.

[The non-trivial automorphism of K/k is complex conjugacy. Show that $\eta \mapsto \eta/\bar{\eta}$ induces a homomorphism $\mathfrak{O}_K^* \to \mu_K/\mu_K^2$ the kernel of which contains $\mathfrak{o}_k^* \mu_K$. Conversely, show that if α is in the kernel then $\alpha/\bar{\alpha} = \zeta^2 = \zeta/\bar{\zeta}$ for some ζ in μ_K; so α/ζ is in $\mathfrak{o}_k^*$.]

1.9 Show that the only integral solutions of $X^3 = Y^2 + 13$ are given by $X = 17, Y = \pm 70$.

[Factorize the equation in $\mathbf{Q}(\sqrt{-13})$, which has class number 2. Show that if x, y is an integer solution then $(y + \sqrt{-13})$ must be the cube of an ideal and hence $y + \sqrt{-13} = (a + b\sqrt{-13})^3$; thus $1 = b(3a^2 - 13b^2)$.]

1.10 Let k be an algebraic number field. Show that there is a finite set of prime ideals $\mathfrak{p}_1, \ldots, \mathfrak{p}_r$ with the following property: if R is the ring consisting of those elements of k which are integers except perhaps at the $\mathfrak{p}_i$, then R is a principal ideal domain.

1.11 Let $\zeta^n = 1$ and assume that $\alpha = (\sum_{i=1}^m \zeta^{n_i})/m$ is an algebraic integer. Show that either $\alpha = \zeta^{n_i}$ for each i or $\alpha = 0$.

1.12 Let X be an indeterminate. Show that the ring $\mathbf{Z}[X]$ is Noetherian and integrally closed in its field of fractions, but is not a Dedekind domain.

1.13 Let k be an algebraic number field. Show that $\mathfrak{o}_k$ is a principal ideal domain if and only if it satisfies the following condition: for every α in k but not in $\mathfrak{o}_k$ there exist β, γ in $\mathfrak{o}_k$ such that

$$0 < |\mathrm{norm}_{k/\mathbf{Q}}(\alpha\beta - \gamma)| < 1.$$

Chapter 2

2.1 If K, k are algebraic number fields with $K \supset k$ prove that the relative discriminant of K/k is $\mathrm{norm}_{K/k}\mathfrak{d}_{K/k}$.

2.2 Show that $X^2 - 82Y^2 = \pm 2$ has solutions in every $\mathbf{Z}_p$ but not in $\mathbf{Z}$. What conclusion can you draw about $\mathbf{Q}(\sqrt{82})$?

2.3 Give an example of finite extensions K_1, K_2 over $\mathbf{Q}$ such that

$$[K_1 K_2 : K_1] \neq [K_2 : K_1 \cap K_2].$$

[Thus without the Galois condition, nothing is left of Lemma 25.]

2.4 Show that the class of the relative discriminant for K/k in the ideal class group $\mathcal{C}_k$ is a square.
[Argue locally, using the fact that $\mathfrak{O}_K \otimes_\mathfrak{o} \mathfrak{o}_\mathfrak{p}$ is a free $\mathfrak{o}_\mathfrak{p}$-module.]

2.5 Let $K \supset k$, let $\mathfrak{P}$ be a prime ideal in K and let $\mathfrak{p}$ be the prime ideal of k divisible by $\mathfrak{P}$. Show that $\mathfrak{P}$ is wildly ramified if and only if $\mathrm{Tr}_{K_\mathfrak{P}/k_\mathfrak{p}}\alpha$ is in $\mathfrak{p}_\mathfrak{p}$ for every α in $\mathfrak{O}_\mathfrak{P}$.

2.6 Let ℓ be a prime and let $f(X) = X^\ell - aX - b$ be an irreducible polynomial in $\mathbf{Z}[X]$ for which $(\ell - 1)a$ and ℓb are coprime. Let K be the splitting field of $f(X)$ over $\mathbf{Q}$. Show that

(i) if p ramifies in $K/\mathbf{Q}$ then $e_p = 2$,
(ii) $G = \mathrm{Gal}(K/\mathbf{Q})$ is S_ℓ, the symmetric group on ℓ elements,
(iii) if k is the fixed field of the alternating group A_ℓ, then K/k is unramified.

[Since $X f'(X) - \ell f(X)$ is linear and not divisible by any prime, any repeated factor of $\tilde{f}(X)$ must be linear and have multiplicity 2. Now Theorem 19 implies (i) and shows that the non-trivial element of the ramification group of p is a transposition. Since G contains an element of order ℓ and a transposition, this gives (ii); and (iii) follows by applying Theorem 16 to the ramification group of p.]

2.7 Prove that if $\mathfrak{p}$ is unramified in K_1/k and in K_2/k, then it is unramified in $K_1 K_2/k$; and that if $\mathfrak{p}$ splits completely in K_1/k and K_2/k then it splits completely in $K_1 K_2/k$. Show however that if $\mathfrak{p}$ remains prime in $K_1 K_2/k$ then $[K_1 K_2 : k]$ is the least common multiple of $[K_1 : k]$ and $[K_2 : k]$.

Give an example where p is ramified in $K_1 K_2/\mathbf{Q}$ but not in $K_1/\mathbf{Q}$ or $K_2/\mathbf{Q}$.

[For the last part, take $p = 3$ and K_1, K_2 to be conjugate fields $\mathbf{Q}(\sqrt[3]{m})$.]

2.8 Let $K = \mathbf{Q}(\sqrt{7}, \sqrt{13})$. Show that for any integer α in K, the discriminant of $1, \alpha, \alpha^2, \alpha^3$ is divisible by 3; and deduce that $1, \alpha, \alpha^2, \alpha^3$ can never be a base for the integers of K.

[Use the fact that (3) splits completely in K, and therefore $3|(\alpha^3 - \alpha)$.]

2.9 Show that $x^4 + 1$ is reducible in $\mathbf{Q}_p$ for every $p > 2$.

Chapter 3

3.1 Let $\rho = \zeta_n + \zeta_n^{-1}$ where ζ_n is a primitive n-th root of unity. What additional condition on n is needed for the ring of integers of $\mathbf{Q}(\rho)$ to be $\mathbf{Z}[\rho]$?

3.2 Let p be a prime and m an integer such that $p|(m^\nu - 1)$ for $\nu = n$ but not for any smaller $\nu > 0$. Show that p has at least one prime factor with $e = f = 1$ in $\mathbf{Q}(\sqrt[n]{1})$ and deduce that $p \equiv 1 \bmod n$. Hence show that for any n there are infinitely many primes $p \equiv 1 \bmod n$.

3.3 Let $K = \mathbf{Q}(\sqrt[n]{1})$ with $n = p_1 \cdots p_m$, where the p_μ are distinct odd primes, and let K_μ be the field of (n/p_μ)-th roots of unity. Let σ_μ be a generator of the cyclic group $\mathrm{Gal}(K/K_\mu)$, write $\sigma = \sigma_1 \cdots \sigma_m$ and let L be the fixed field of σ. Show that K/L is unramified at each finite place. What additional condition is needed to ensure that it is also unramified at the infinite places?

3.4 (i) Let G be a finite abelian group. Show that there are fields K, L with $K = \mathbf{Q}(\sqrt[n]{1})$ and $L \subset K$ such that $\mathrm{Gal}(K/L) \approx G$ and no (finite or infinite) place ramifies in K/L. [Use the results of the two previous exercises.]

(ii) Using Theorem 37, show that the ideal class group of L contains a subgroup isomorphic to G.

The following two exercises illustrate the method of infinite descent, and show the importance of choosing the right equation to apply the method to.

3.5 Show that $X^4 + Y^4 = Z^2$ has no non-trivial solutions in $\mathbf{Z}$.

[If there are solutions, let x, y, z be the one with $|z|$ minimal. Without loss of generality, assume y odd; then $x^2 = 2uv$, $y^2 = u^2 - v^2$ with u, v coprime. Thus $u = A^2$, $v = 2B^2$ and $y^2 = A^4 - 4B^4$. Hence show that $A^2 = \ell^2 + m^2$, $B^2 = \ell m$ with ℓ, m coprime and therefore both squares. Now $|A| < |z|$ gives a contradiction.]

3.6 Show that $\epsilon X^4 + Y^4 = Z^2$ has no non-trivial solutions in $\mathbf{Z}[\sqrt{-1}]$ where $\epsilon^4 = 1$.

[Let $\pi = 1 + \sqrt{-1}$. Show by π-adic arguments that it is enough to consider the case when π divides X but not Y or Z, and that then $\pi^2 | X$. Deduce that we can take $Z - Y^2 = \pi^2 \epsilon_1 u^4$, $Z + Y^2 = \pi^{-2} \epsilon_2 v^4$ where ϵ_1, ϵ_2 are roots of unity; and hence $Y^2 = \pi^{-4} \epsilon_3 v^4 + \epsilon_4 u^4$ where $\pi^2 | v$. Writing $Y\sqrt{-1}$ for Y if necessary, show that $\epsilon_4 = 1$. Verify that the same power of π divides v and X, and derive a contradiction.]

3.7 Repeat the investigation at the end of §14 for the case when K is the splitting field of an irreducible non-normal cubic equation over $\mathbf{Q}$.

3.8 Verify equation (46) by elementary means.

[The only substantial calculation concerns the powers of 2; here it pays to split cases, using the properties both of the different and of a base for $\mathfrak{O}_K$.]

3.9 Let $k = k_m$ be the field of m-th roots of unity and k_0 its maximal totally real subfield; and let μ_k be the group of roots of unity in k.

(i) If m is a prime power or twice a prime power, show that $\mathfrak{o}_k^* = \mathfrak{o}_{k_0}^* \mu_k$.
(ii) If not, show that $[\mathfrak{o}_k^* : \mathfrak{o}_{k_0}^* \mu_k] = 2$.

[Use the result of Example 1.8. For (i) let $m = p^r$ or $2p^r$. If the result is false and $p > 2$, there exists η in $\mathfrak{o}_k^*$ with η^2 in k_0 but η not in k_0; then $k = k_0(\sqrt{\eta})$ would not be totally ramified at p. If $p = 2$ and η in $\mathfrak{o}_k^*$ is not in $\mathfrak{o}_{k_0}^* \mu_k$ then $\zeta = \eta/\bar{\eta}$ is a primitive m-th root of unity; then $\text{norm}_{k/L}\zeta = \pm\sqrt{-1}$ but $\text{norm}_{k/L}\eta$ is a power of $\sqrt{-1}$ where $L = \mathbf{Q}(\sqrt{-1})$. For (ii), show that $\eta = 1 - \zeta$ is a unit, where ζ is a primitive m-th root of unity and deduce that the homomorphism of Exercise 1.8 is onto.]

Chapter 4

4.1 If the Dirichlet series $\sum a_n n^{-s}$ converges at $s = s_0$, show that it converges whenever $\Re s > \Re s_0$ and that it converges absolutely whenever $\Re s > \Re s_0 + 1$.

[For the first assertion, use summation by parts.]

4.2 If the limit (98) exists, show that the limit (99) exists and has the same value.

[Use summation by parts.]

4.3 Let $\mathcal{S}$ be a finite set of places and let χ be a character on $I^{\mathcal{S}}$ such that for any neighbourhood $\mathcal{N}$ of 1 there exists $\epsilon > 0$ such that the inverse image of $\mathcal{N}$ contains all principal ideals (ξ) with $\|\xi - 1\|_v < \epsilon$ for each v in $\mathcal{S}$. Show that χ satisfies (91).

[Choose the n_v, s_v so that (91) holds for all units with $\|\xi - 1\|_v < \epsilon$; then proceed as in the proof of Lemma 34.]

4.4 Let χ be a non-trivial character of conductor m; show that

$$\left| \sum_{n=1}^{N} \chi(n) \right| \leqslant m^{1/2}(1 + \log m)$$

for all integers N. Deduce that if p is prime and χ is a non-trivial character $\bmod\, p$ then $\chi(n) \neq 1$ for some $n \leqslant p^{1/2}(1 + \log p)$.

[If $\tau_x(\chi)$ is the Gauss sum as defined in (54), show that

$$\sum_{1}^{N} \chi(n) = (\tau_1(\overline{\chi}))^{-1} \sum_{n=1}^{N} \sum_{b \bmod m} \overline{\chi}(b) \exp(2\pi i b n / m)$$

$$= (\tau_1(\overline{\chi}))^{-1} \sum_{b \bmod m} \frac{\exp\{2\pi i b (N+1)/m\} - 1}{\exp\{2\pi i b / m\} - 1}.$$

In each term of the last sum, the numerator is absolutely bounded by 2 and the absolute value of the denominator is $|2\sin(\pi b/m)|$. Hence

$$\left| \sum \chi(n) \right| \leqslant 2m^{-1/2} \sum \operatorname{cosec}(\pi b/m)$$

where we can take the sum to run over $0 < b \leq \frac{1}{2}m$.]

Chapter 5

5.1 Write $i = \sqrt{-1}$ and let $K = \mathbf{Q}(i, \sqrt[4]{2})$, $k_1 = \mathbf{Q}(i)$ and $k_2 = \mathbf{Q}(\sqrt{2})$. Show that K is abelian over each of k_1 and k_2, with $\mathrm{Gal}(K/k_1) \approx C_4$ and $\mathrm{Gal}(K/k_2) \approx C_2 \times C_2$.

(i) If H is the congruence divisor class associated with K/k_1, show that its conductor divides $(1 + i)^7$ and that H contains i and 5, and that these facts specify it completely.

(ii) If H is the congruence divisor class associated with K/k_2, show similarly that the finite part of its conductor divides $(\sqrt{2})^5$ and that H contains 3, and that these facts specify it completely.

[In (i), every $\alpha \equiv 1 \bmod (1+i)^7$ is a $(1+i)$-adic fourth power and $(1+i)$ is the only prime which ramifies in K/k_1; these give the assertion about the conductor. Since the units represent all the odd residue classes $\bmod (1+i)^3$, we need only identify the residue classes $\bmod (1+i)^7$ in H which are congruent to $1 \bmod (1+i)^3$. This appears to give six candidates for H, but four can be rejected because they are not invariant under complex conjugacy. One of the remaining two contains 5 and the other contains $1 + 4i$; but the second must be rejected because $(1 + 4i)$ does not split completely in K/k_1.]

5.2 Let $X_1, \dots, X_4$ be integers not all divisible by any $p \equiv 7 \bmod 8$, such that

$$X_1^4 + 4X_2^4 = (X_1^2 - 2X_1X_2 + 2X_2^2)(X_1^2 + 2X_1X_2 + 2X_2^2) = X_3^2 - 2X_4^2.$$

Show that $|X_3| \not\equiv 5$ or $7 \bmod 8$.

[Assume that X_3 is odd and show that $4|X_2$. Using the notation and results of the previous exercise, show successively that the ideal $(X_1^2 + 2iX_2^2)$ is in the kernel of the Artin reciprocity map for K/k_1, that the first degree primes in this kernel are those whose Norms split completely in K, and that therefore the ideal $(X_3 + X_4\sqrt{2})$ is in the kernel of the Artin reciprocity map for K/k_2. Hence deduce that X_4 is even and $|X_3| \equiv 1$ or $3 \bmod 8$.]

5.3 Let k be totally complex and k_0 totally real with $[k : k_0] = 2$. If h, h_0 are the class numbers of k, k_0 respectively, show that $h_0 | h$.

[Let L be the Hilbert class field of k_0. Show that $L \cap k = k_0$ whence $[kL : k] = [L : k_0] = h_0$, and that kL/k is unramified and abelian.]

5.4 Let $\left(\frac{m}{n} \right)$ denote the quadratic residue symbol in $\mathbf{Z}$, where m, n are coprime and n is odd and positive. Using the general properties of the power residue symbol, show that $\left(\frac{-1}{n}\right)$ and $\left(\frac{2}{n}\right)$ only depend on $n \bmod 8$, and hence deduce the auxiliary laws

$$\left(\tfrac{-1}{n}\right) = (-1)^{(n-1)/2}, \quad \left(\tfrac{2}{n}\right) = (-1)^{(n^2-1)/8}.$$

It is possible to derive Theorem 25 in the same way, but this involves an unnatural case-by-case calculation; a somewhat less ugly version of the same calculation can be found in the next exercise.

5.5 (i) Let $m = 2$ and $k = \mathbf{Q}$ in the machinery of §19. Show that the Hilbert symbol $(a, b)_2$ depends only on a and $b \bmod (\mathbf{Q}_2^*)^2$, and construct a table of its values. [Choose particular representatives for the classes of a and b and use (104) and Lemma 38.]

(ii) Check the results of (i) by means of (105).

(iii) Deduce the law of quadratic reciprocity for $\mathbf{Q}$.

5.6 Take $p = 3$ in the notation of the exercise on page 111. Show that any α in k^* can be written in essentially one way in the form $\alpha = \zeta^\mu \pi^\nu \alpha_0$ with $\alpha_0 \equiv \pm 1 \bmod 3$. Prove that

$$\left(\frac{\alpha_0}{\beta_0}\right) = \left(\frac{\beta_0}{\alpha_0}\right)$$

if each of α_0, β_0 is congruent to $\pm 1 \bmod 3$ and $S(\alpha_0) \cap S(\beta_0) = S$. Prove also that

$$\left(\frac{\zeta}{\alpha_0}\right) = \zeta^{-r-s}, \quad \left(\frac{\pi}{\alpha_0}\right) = \zeta^r$$

if $\alpha_0 = \pm(1 + 3(r + s\zeta))$.

Suggested further reading

There are a substantial number of texts which cover more or less the same material as Chapters 1 and 2. The most thorough and careful is
[FT] A. Fröhlich and M.J. Taylor, *Algebraic Number Theory*, Cambridge University Press, 1991.
The texts which I have found most inspiring are
[Weyl] H. Weyl, *Algebraic Theory of Numbers*, Princeton University Press, 1940,
[Ca] J.W.S. Cassels, *Local Fields*, Cambridge University Press, 1986. This goes far wider than its title suggests, with a particularly strong emphasis on applications to Diophantine equations.
A more comprehensive account of the local theory, both elementary and advanced, is
[Se] J.-P. Serre, *Local Fields*, Springer-Verlag, 1979. (The original French version was *Corps Locaux*, Hermann, 1968.) This can fairly be described as a masterpiece.

The best reference for class field theory (covering both local and global fields) is still
[CF] J.W.S. Cassels and A. Fröhlich (editors), *Algebraic Number Theory*, Academic Press, 1967.
Alternative accounts can be found in
[N] J. Neukirch, *Class Field Theory*, Springer-Verlag, 1986;
[L] S. Lang, *Algebraic Number Theory*, Springer-Verlag, 1986. This was originally written in 1970; it also covers the elementary theory and a substantial amount of analytic material.
An interesting and recommendable illustration of how much can be achieved without overt Galois cohomology is the second half of
[Weil] A. Weil, *Basic Number Theory*, Springer-Verlag, 1973. The first half

of this book is an account of the elementary theory, dominated by the use
of Haar measure; this is an interesting approach, but Weil's style does not
help the reader.

The best account of Iwasawa theory is to be found in

[Wa] L. Washington, *Introduction to Cyclotomic Fields*, Springer-Verlag,
1980.

Books on the computational aspects of the subject become out of date as
new packages are developed. Subject to this, the definitive account is

[Co] H. Cohen, *A Course in Computational Algebraic Number Theory*,
Springer-Verlag, 1995.

A good introduction to these aspects can be found in

[Sm] N.P. Smart, *The Algorithmic Resolution of Diophantine Equations*,
Cambridge University Press, 1998.

Index

d_k, 4
h, 15
J_k, 15, 50
$k_{\mathfrak{p}}$, 34
R, 23
r_1, r_2, 3
V_k, 49
w, 53
$\mathbf{T}$, 125
$\mathfrak{o}, \mathfrak{O}$, 1
Δ^2, 4
$\mathcal{C}_k$, 15
$\mathcal{S}_k$, 16

absolute norm, 13
absolute value, 31
adèle, 48
    principal, 49
algebraic number field, vii
approximation
    strong, 40
    weak, 39
Archimedean, 32
Artin element, symbol, 29
Artin map, 100
    local, 108
Artin Reciprocity Law, 100
ascending chain condition, 6

Cebotarev Density Theorem, 96
character, 125
    congruence, 79
    Hecke Grössencharakter, 82, 90, 104
    Tate, 84
characteristic polynomial, 121
Chinese Remainder Theorem, 12
circle group, 125
class field, 99
class field theory, viii
class number, 15

conductor, 43, 68, 99, 100
congruence divisor class group, 99
conorm, 23
cyclotomic, 65

Dedekind domain, 9
density, 94
    Dirichlet, 94
diagonal map, 49, 51
different, 43, 44
    local, 44
discriminant
    absolute, 4
    relative, 26

Exercise, 17, 19, 40, 43, 62, 69, 76, 111

Fermat's Last Theorem, vii, 73
first degree primes, 94
Frobenius element, 29, 134

Gauss sum, 69

Haar measure, integral, 123
Hasse Norm Theorem, 102
height, 16
Hensel's Lemma, 35
Hilbert Basis Theorem, 8
Hilbert symbol, 107, 108
Hilbert's Theorem 90, 4

ideal
    fractional, 10
    integral, 10
ideal class group, 15
idèle, 48, 50
    principal, 51
idèle class group, 105
inertia group, field, 28
integer, algebraic, 1

145

integral at p, 14
integral closure, 3

Kronecker-Weber Theorem, viii, 112

L-series, Dirichlet, 79
lattice, 120
local field, 35

minimal base, 118

Noetherian, 6
norm, 121

order, 3

pigeonhole principle, 16, 120
place, 31
    finite, 34
    infinite, 33
Pontryagin duality, 125
power residue symbol, 106
Product Formula, 35, 50

Quadratic Reciprocity Law, 61
quasi-character, 128

ramification, 41
    tame, 41
    wild, 41
ramification group, 28
ramified primes, 15
reduced form, 57
regulator, 23
Riemann hypothesis, 79

splitting group, field, 27
Stickelberger, 5

trace, 121
transform
    Fourier, 128
    Mellin, 130

ultrametric, 34
unit, 15
    cyclotomic, 71
    fundamental, 58

valuation
    additive, 35
    multiplicative, 31

zeta function
    Riemann, 79
    Tate, 84